T0337838

# Processing of Nanoparticle Structures and Composites

# Processing of Nanoparticle Structures and Composites

## Ceramic Transactions, Volume 208

*A Collection of Papers Presented at the
2008 Materials Science and Technology
Conference (MS&T08)
October 5–9, 2008
Pittsburgh, Pennsylvania*

Edited by

Tom Hinklin
Kathy Lu

**A John Wiley & Sons, Inc., Publication**

Published by John Wiley & Sons, Inc., Hoboken, New Jersey.
Published simultaneously in Canada.

For general information on our other products and services or for technical support, please contact our Customer Care Department within the United States at (800) 762-2974, outside the United States at (317) 572-3993 or fax (317) 572-4002.

Wiley also publishes its books in a variety of electronic formats. Some content that appears in print may not be available in electronic format. For information about Wiley products, visit our web site at www.wiley.com.

*Library of Congress Cataloging-in-Publication Data is available.*

ISBN 978-0-470-40846-9

Printed in the United States of America.

10 9 8 7 6 5 4 3 2 1

# Contents

# Introduction

This volume contains a collection of papers from the Controlled Processing of Nanoparticle Structures and Composites symposia held during the 2008 Materials Science and Technology conference (MS&T08)—a joint meeting between ACerS, AIST, ASM International, and TMS at the David L. Lawrence Convention Center, Pittsburgh, Pennsylvania, USA, October 5–9, 2008.

# NANOPARTICLE-BASED BULK MATERIAL TEMPLATING

Kathy Lu,* Chase Hammond
Materials Science and Engineering Department
Virginia Polytechnic Institute and State University
Blacksburg, VA 24061, USA

ABSTRACT
        Nanoparticle size decrease has opened a whole new field for unique particle-based materials forming. This study is focused on making molds with micron-size feature arrays and freeze casting of surface templated materials. Polydimethylsiloxane (PDMS) and silicone molds with micron-size feature arrays are made using a mold core templated by a focused ion beam. Templated surface feature transfer ability is compared for the PDMS and silicone molds. The process can make complex, near-net shape, and fine feature arrays. This work provides a new approach for surface templating of nanoparticle based materials.

## INTRODUCTION

        Templating is a key enabler for fulfilling the promise of nanotechnology.[1-3] It capitalizes on the special material properties and processing capabilities at the nanoscale, and promotes the integration of nanostructures to multifunctional micro-devices and meso-/macro-scale architectures and systems, as well as the interfaces across multi-length scales. Nanoparticle-based processing has been an active research area in this regard because of the potential to create large surface area materials and integrate heterogeneous structures. However, most current efforts are either focused on nanoparticle suspensions or nanoparticle-based thin film structures. The barriers for developing bulk nanostructured materials are agglomeration, inhomogeneous microstructures, and lack of structural control across multi-length scales.

        More recently, templated materials containing intricate, nanometer-scale structures have been developed. They frequently exhibit novel, highly anisotropic, or enhanced properties, directly related to the dimensionality and the extra degrees of freedom of the nanostructure.[4-6] In our prior work, substantial efforts have been devoted to developing high solids loading nanoparticle suspensions.[7-11] This development opens up numerous opportunities for templating nanoparticle-based bulk materials. In addition, the small size of the nanoparticles offers the potential to engineer nanostructured materials to produce unique features in large surface areas.

        Based on our knowledge in making high solids loading nanoparticle suspensions, a templating approach is used to make nanoparticle-based materials. A focused ion beam (FIB) based process is pursued to produce the designed features on silicon wafer first. Then polymeric molds are made using the FIB templated silicon wafer. High solids loading $Al_2O_3$ nanoparticle suspension is cast into the polymeric molds and converted into solid state by freeze casting. The work shows a new approach of producing micron-size feature arrays on the surface of $Al_2O_3$ nanoparticle materials.

## EXPERIMENTAL PROCEDURE

### Template Making

        The starting template for making the molds was obtained using a dual beam FIB instrument (FEI Helios 600 NanoLab, Hillsboro, OR). The instrument is composed of a subnanometer resolution field emission scanning electron microscope (SEM) and a field emission scanning Ga$^+$ beam column. The specimen used in templating can be moved by 150 mm distance along X and Y axes and tilted from -5 to 60° by a high precision specimen goniometer. The Ga$^+$ source has a continuously adjustable energy range from 0.5 kV to 30 kV, and an ion current between 1.5 pA and 21 nA. The patterning beam spot size and resolution vary with the ion beam current and voltage and can be tuned to as small as 5 nm. The FIB instrument also has a high resolution, 24 bit digital patterning engine capable of simultaneous patterning and imaging. Si wafer (p type, (100) orientation) was used as the starting template material for making the polymeric molds.

Mold Making

PDMS ((H$_3$C)$_3$Si[Si(CH$_3$)$_2$O]$_n$Si(CH$_3$)$_3$, Dow Corning Corporation Midland, MI) and silicone (RTV 664, General Electric Company, Waterford, NY) were used as the polymeric mold making materials with the templated silicon wafer as the mold core. Silicone molds without the templated features have been produced before.[7-11] In this study, silicone base compounds and the corresponding curing agent were homogeneously mixed at 10:1 ratio before being poured over the silicon wafer. After that the silicon wafer and the silicone mixture were placed under 40 Pa pressure for air removal. Since the silicone mixture had high viscosity, the vacuuming process was repeated three to five times before being kept at 40 Pa for 15 minutes. Then the molds were cured for 24 hrs in air. The PDMS molds were made using a similar process. PDMS base compounds and the corresponding curing agent were mixed at 10:1 ratio. The mixture proved much less viscous than the silicone system. It was poured over the silicon wafer and allowed to sit for a few minutes. The silicon wafer and the PDMS were then placed in a vacuum chamber and subjected to a pressure of approximately 27 Pa. This effectively removed all porosity resulting from the air bubbles inherent in the PDMS mixture. Once all air had been removed the silicon wafer and PDMS were placed into a drying oven at 100°C. After 45 minutes the solidified PDMS mold was separated from the silicon wafer.

Al$_2$O$_3$ Suspension Making

Al$_2$O$_3$ nanoparticles with specific surface area of 45 m$^2$/g were used in this study (Nanophase Technologies, Romeoville, IL). Average Al$_2$O$_3$ particle size was around 38 nm with a normal size distribution from 10 to 50 nm.[7] For the Al$_2$O$_3$ nanoparticle suspension preparation, poly(acrylic acid) (PAA, M$_W$ 1,800, Aldrich, St Louis, MO) was used as a dispersant with the segment as [-CH$_2$CH(CO$_2$H)-]; glycerol (C$_3$H$_8$O$_3$, Fisher Chemicals, Fairlawn, NJ) was used as part of the dispersing medium with the molecular formula as CH$_2$OH-CHOH-CH$_2$OH. Water-glycerol mixture at a ratio of 9:1 (water: glycerol) was used as the dispersing medium. The mixture was homogenized for 5 minutes using a ball mill before use. Al$_2$O$_3$ nanoparticles were added into the dispersing medium in 10 g increments along with 2.0 wt% of PAA dispersant (on Al$_2$O$_3$ basis). Since low pH promotes PAA dispersant adsorption onto Al$_2$O$_3$ nanoparticles,[12] HCl solution was added to lower the pH to 1.5. The suspension was ball milled for 12 hrs with periodic adjustment of pH to 1.5. Suspensions of approximately 20 vol% Al$_2$O$_3$ solids loading were made by this procedure. After this step, Al$_2$O$_3$ nanoparticles were again added in 10 g increments, along with 2.0 wt% of PAA dispersant (on Al$_2$O$_3$ basis) to make 35 vol% solids loading suspension. NH$_4$OH was used to adjust the suspension to pH 9.5. The suspension was then mixed for 24 hrs for complete homogenization.

Templating

Al₂O₃ nanoparticle suspension was filled into the silicone and PDMS molds immediately after the suspension preparation using a disposable pipette. Care was taken to completely fill the molds and avoid air bubbles. The filled molds were kept under ambient conditions for 1 hr. After this pre-rest, the samples were frozen in a freeze dryer (AdVantage El-53, SP Industries Inc., Warminster, PA) immediately. The freezing rate used was 0.25°C/min. The freezing temperature was -35°C. At the freezing temperature, the samples were kept for 2 hrs before the chamber pressure was decreased to 1 Pa. The filled molds were kept at -35°C and 1 Pa pressure for 10 hrs and then heated to room temperature in stages (-20°C for 8 hrs, -10°C for 4 hrs, -5°C for 5 hrs, and 5°C for 5 hrs). During the entire process, a pressure of 1 Pa was maintained.

The templated surfaces of the freeze dried Al₂O₃ nanoparticle samples were analyzed using a LEO550 field emission SEM (Carl Zeiss MicroImaging, Inc, Thornwood, NY). To illustrate the three dimensional nature of the templated features, all the SEM images were taken at 45° tilt angle.

RESULTS AND DISCUSSION

Templated Silicon Mold Core

To template micron size features of different shape and size from the Al₂O₃ nanoparticle suspension by freeze casting, two steps are needed in making the freeze casting molds. The first step is to make the mold core with a designed pattern. The second step is to make the molds that can confine the Al₂O₃ nanoparticle suspension and produce solid state Al₂O₃ samples with fine features. As mentioned, a silicon wafer is used as the mold core and the designs to be produced on the silicon wafer are shown in Figure 1. There are three different patterns: homogenous circle arrays, heterogeneous circle arrays, and homogenous square arrays. Each array size is 100x100 μm. The larger size circles are 10 μm in diameter, the small size circles are 5 μm in diameter, and the squares are 10 μm in edge length. The feature-to-feature distance is 15 μm. The features at the array edge are 20 μm from the designed pattern boundary. During the FIB patterning, the white area material is removed. The black area material is left intact.

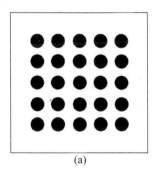
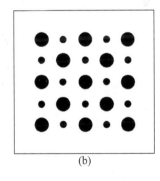

| (a) | (b) |

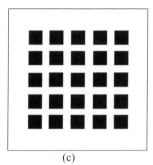

(c)

Figure 1. Feature array designs for FIB patterning: (a) homogenous circle arrays, (b) heterogeneous circle arrays, and (c) homogenous square arrays.

Under 30 kV voltage and 21 nA beam current patterning condition, the patterns created are shown in Figure 2. As shown, the arrays have well-defined size, shape, and feature arrangement, as well as smooth surfaces. The patterning time is approximately 70 min for the homogenous round and square island arrays and 30 min for the heterogeneous round island arrays. The feature height is controlled by the patterning time. The homogenous round and square island array height is approximately 4.8-5.0 μm, and the heterogeneous round island array height is approximately 2.0-2.5 μm. One observation is that it is important to obtain well focused ion beam for patterning. Good beam focus produces smoother feature surfaces and the same pattern in a faster rate.[13]

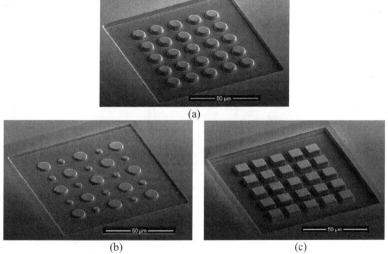

Figure 2. Feature arrays produced from FIB patterning: (a) homogenous round island arrays, (b) heterogeneous round island arrays, and (c) homogenous square island arrays.

PDMS and Silicone Molds

The PDMS and silicone molds are made from the silicon mold core with the feature arrays shown in Figure 2. Figure 3 shows the SEM images of the surface of the PDMS mold with the features present. Figure 4 shows the SEM images of the surface of the silicone mold with the features present. For both polymeric molds, the feature arrays are accurately produced in an inverse manner. The new cavity features are well defined and free of porosity. The depth of the heterogeneous round cavity features is roughly half of those of the homogeneous round and square cavity features. The large flaw in one of the square features in Figures 3(c) and 4(c) is a result of a chip-off on the silicon wafer, demonstrating the feature array's reproducibility.

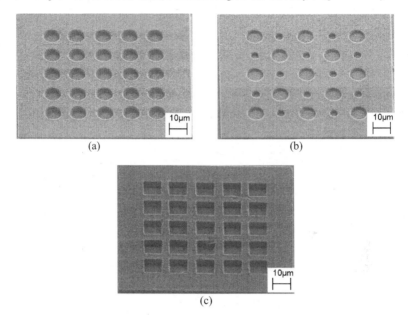

(a)

(b)

(c)

Figure 3. Cavity feature arrays produced on PDMS mold using FIB patterned silicon wafer core: (a) homogenous round cavity array, (b) heterogeneous round cavity array, and (c) homogenous square cavity array.

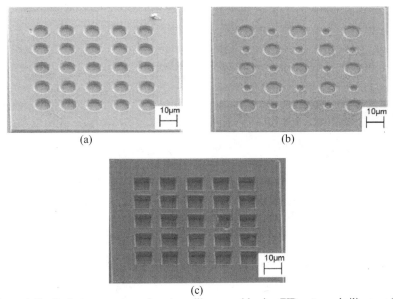

(a)                                              (b)

(c)

Figure 4. Cavity feature arrays produced on silicone mold using FIB patterned silicon wafer core: (a) homogenous round cavity array, (b) heterogeneous round cavity array, and (c) homogenous square cavity array.

Even though both PDMS and silicone mold materials produce equally desirable and defined feature shapes and arrangement, the molds made from the two materials have different characteristics. The silicone mold is more rigid than the PDMS mold at room temperature. Also, the silicone mold has opaque blue color while the PDMS mold is transparent. The different mold rigidity and transparency are products of the inherent chemical composition differences between the two materials.

Templated Al$_2$O$_3$ Nanoparticle Samples

Figure 5 shows the SEM images of freeze dried Al$_2$O$_3$ nanoparticle feature arrays produced with the PDMS mold. Figure 6 shows the SEM images of freeze dried Al$_2$O$_3$ nanoparticle feature arrays produced with the silicone mold. As it shows, the feature arrays are reproduced on the surface of the freeze-dried Al$_2$O$_3$ nanoparticle bulk samples. However, there is a clear feature reproducibility difference. The PDMS mold results in better defined features than the silicone mold for the Al$_2$O$_3$ nanoparticle samples. The freeze cast samples from the PDMS mold have sharper corners and edges, precise arrangement, and uniform height. There are some cracks and fracture for individual features, but the arrays are generally well controlled. This means the compliant PDMS mold becomes more rigid under the freeze casting condition, which is conducive for feature shape retention. In contrast, the feature arrays from the silicone mold, through, are severely deformed; there is substantial feature-to-feature difference. As to be shown in Figure 7, the silicone molds with gold-platinum coating can reproduce the designed feature arrays under freezing casting conditions. This means the silicone mold can maintain its rigidity and templated mold shapes at the freezing condition. The main cause for the poor feature reproducibility results from the mold surface affinity with the $Al_2O_3$ nanoparticle suspension. The sticking of the $Al_2O_3$ nanoparticle features to the silicone mold surface leads to the severe distortion of the features. This aspect will be further quantified in future studies.

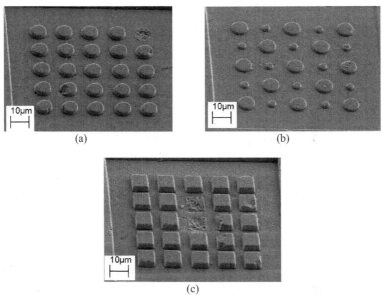

(a)  (b)

(c)

Figure 5. $Al_2O_3$ nanoparticle arrays produced from PDMS mold: (a) homogenous round island array, (b) heterogeneous round island array, and (c) homogenous square island array.

(a)  (b)

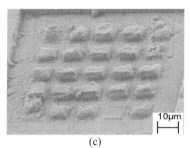

(c)

Figure 6. Al$_2$O$_3$ nanoparticle arrays produced from silicone mold: (a) homogenous round island array, (b) heterogeneous round island array, and (c) homogenous square island array.

The formation of bulk, solid state Al$_2$O$_3$ nanoparticle samples is sensitive to the freeze casting conditions.[14] A pre-rest of an hour has to be used in accordance with previous work in order to effectively remove the air bubbles trapped inside the Al$_2$O$_3$ nanoparticle suspension.[7-11] Freezing rates, drying rates, and drying temperatures need to be carefully controlled. Freezing rate is the most significant parameter affecting the density and integrity of the freeze cast samples. Fast freezing rates result in an increase in the porosity of the templated Al$_2$O$_3$ nanoparticle samples and substantially lower the sample strength. Slow freezing rates can reduce the porosity and improve the strength of the Al$_2$O$_3$ nanoparticle sample. The freezing rate for the 35 vol% solids loading Al$_2$O$_3$ suspension needs to be 0.25°C/min or less. The time and temperature of the drying steps have a smaller effect on the templated Al$_2$O$_3$ nanoparticle samples, but still affect the final results. The current drying time and temperature seem to be too elaborate and can be further simplified.

The fundamental cause for the cracking and fracture of the features seen in Figures 5 and 6 is the higher than desired interfacial energy, or affinity, between the molds and the Al$_2$O$_3$ nanoparticle samples. Coating the molds with a gold-platinum layer shows to be an effective approach to reducing the feature cracking and damage. For example, some silicone molds have been coated with a 10 nm gold-platinum layer. Figure 7 shows the corresponding freeze cast feature arrays. There are almost no cracks or fractures on the features, most noticeably for the square features. Again the defect due to the chip-off from the silicon wafer is reproduced. The gold-platinum nanolayer appears to decrease the adherence between the molds and the Al$_2$O$_3$ nanoparticle samples and result in less feature fractures and cleaner mold. However, the gold-platinum layer should not be rubbed or scratched in order to avoid nanolayer damage. Wettability of the Al$_2$O$_3$ nanoparticle suspension vs. different mold surfaces is being investigated in our studies.

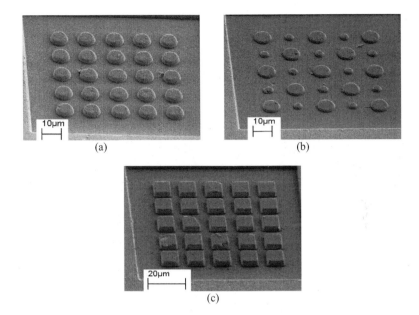

Figure 7. $Al_2O_3$ nanoparticle patterns produced from silicone mold coated with 10 nm thickness gold-platinum layer: (a) homogenous round island array, (b) heterogeneous round island array, and (c) homogenous square island array.

Throughout the templating process, the cleanliness of the mold is crucial for producing well defined features. Many of the defects seen in the samples can be traced back to what is present in the dirty mold. Mold cleaning with distilled water and ethanol tends to leave residue behind. A concrete cleaning procedure would improve the templating process. Coating the mold with a fresh gold-platinum nanolayer seems to simultaneously address this issue.

CONCLUSION
A nanoparticle-based templating process has been studied using $Al_2O_3$ nanoparticle suspension. With templated silicon core to make PDMS and silicone molds and high solids loading $Al_2O_3$ nanoparticle suspensions of good flowability to fill the molds, micron-size arrays of different size and shape islands have been created on $Al_2O_3$ nanoparticle material surfaces. The PDMS molds produce more desirable features under the reported freeze casting condition. Coating a 10 nm thickness gold-platinum layer on the silicone mold surface enables creation of well defined $Al_2O_3$ nanoparticle feature arrays. This templating process can be applied to large surface area nanoparticle-based materials and poses great potentials for direct device fabrication.

ACKNOWLEDGMENT
The authors acknowledge the financial support from National Science Foundation under grant No. CMMI-0824741.

FOOTNOTE
Member, American Ceramic Society

REFERENCES
[1]J. Lu, S. S. Yi, T. Kopley, C. Qian, J. Liu, and E. Gulari, Fabrication of Ordered Catalytically Active Nanoparticles Derived from Block Copolymer Micelle Templates for Controllable Synthesis of Single-Walled Carbon Nanotubes, *J. Phys. Chem. B*, **110**, 6655-6660 (2006).
[2]J. Gierak, E. Bourhis, A. Madouri, M. Strassner, I. Sagnes, S. Bouchoule, M. N. Mérat Combes, D. Mailly, P. Hawkes, R. Jede, L. Bardotti, B. Prével, A. Hannour, P. Mélinon, A. Perez, J. Ferré, J.-P. Jamet, A. Mougin, C. Chappert, and V. Mathet, Exploration of the Ultimate Patterning Potential of Focused Ion Beams, *J. Microlith., Microfab., Microsyst.*, **5**, 011011-011011-011011 (2006).
[3]T. Brezesinski, M. Groenewolt, A. Gibaud, N. Pinna, M. Antonietti, and B. M. Smarsly, Evaporation-Induced Self-Assembly (EISA) at its Limit: Ultrathin, Crystalline Patterns by Templating of Micellar Monolayers, *Adv. Mater.*, **18**, 2260-2263 (2006).
[4]A. Huczko, Template-Based Synthesis of Nanomaterials, *Appl. Phys.*, **70**, 365-376 (2000).
[5]T. L. Wen, J. Zhang, T. P. Chou, S. J. Limmer, and G. Z. Cao, Template-Based Growth of Oxide Nanorod Arrays by Centrifugation, *J. Sol-Gel Sci. Technol.*, **33**, 193–200 (2005).
[6]T. L. Wade and J.-E. Wegrowe, Template Synthesis of Nanomaterials, *Eur. Phys. J. Appl. Phys.*, **29**, 3-22 (2005).
[7]K. Lu, X. Zhu, Freeze Casting as a Nanoparticle Material Forming Method, *Int. J. Appl. Ceram. Technol.*, **5**, 219–227 (2008).
[8]K. Lu, Freeze Cast Carbon Nanotube-Alumina Nanoparticle Green Composites, *J. Mater. Sci.*, **43**, 652-659 (2008).
[9]K. Lu, Microstructural Evolution of Nanoparticle Aqueous Colloidal Suspensions During Freeze Casting, *J. Am. Ceram. Soc.*, **90**, 3753-3758 (2007).
[10]K. Lu, C. S. Kessler, R. M. Davis, Optimization of a Nanoparticle Suspension for Freeze Casting, *J. Am. Ceram. Soc.*, **89**, 2459-2465 (2006).
[11]K. Lu, C. S. Kessler, Nanoparticle Colloidal Suspension Optimization and Freeze-Cast Forming, Ceramic Engineering and Science Proceedings, Synthesis and Processing of Nanostructured Materials, **27**, 1-10, 2006. Ed. W. M. Mullins, A. Wereszczak, and E. Lara-Curzio, Proceeding of 30th International Conference on Advanced Ceramics and Composite, American Ceramic Society, Cocoa Beach, FL.
[12]J. Cesarano and I. A. Aksay, Processing of Highly Concentrated Aqueous Alpha-Alumina Suspensions Stabilized with Poly-Electrolytes, *J. Am. Ceram. Soc.*, **71**, 1062-1067 (1988).
[13]K. Lu, Hierarchical and Nanosized Pattern Formation Using Dual Beam Focused Ion Beam Microscope, *J. Nanosci. Nanotechnol.*, submitted.
[14]K. Lu, C. Hammond, Nanoparticle-based Surface Templating, *Int. J. Appl. Ceram. Technol.*, submitted.

# CONTROLLING THE PROCESSING PARAMETERS FOR CONSOLIDATION OF NANOPOWDERS INTO BULK NANOSTRUCTURED MATERIAL

A. Sadek, H. G. Salem,
Department of Mechanical Engineering, Youssef Jameel Science and Technology Research Center, American University in Cairo-Egypt
hgsalem@aucegypt.edu

## ABSTRACT

Nanostructured materials have attracted many researchers due to their high potential to provide outstanding mechanical and physical properties. In the current research work, powder metallurgy technique was employed in an attempt to produce bulk nanostructured materials (BNSM). AA2124 nanopowders <100nm in particle and 20nm grain size fabricated by high energy ball milling of gas atomized micronpowders was employed. The nanopowders were hot compacted in a confined die using uniaxial single sided pressing for preliminary densification at 0.7Tm of the alloy. Selected intact hot compacts (HCs) were promoted for severe plastic deformation via equal channel angular pressing (ECAP) at the minimum possible deforming temperature (245°C) for final densification. The effect of ECAP single pass and two passes route (A) on the relative density, density uniformity hardness and compressive properties was investigated. Structural evolution during the various stages of processing was inspected by optical and electron microscopy. Bulk nanostructured rods 50-60nm in size with superior properties were produced by combined processing of HC and ECAP. The processed bulk products demonstrate properties equivalent to dispersion strengthened AA2124 metal matrix composite reinforced with 10-15% $Al_2O_3$. 73% of the total grain coarsening occurred during the HC stage, while only 27% grain coarsening took place at the end of 2-pass warm ECAP. The compressive strength dropped by 32% after one pass ECAP due to partial rigid body rotation of the unconsolidated powder particles inherited by the HC, followed by 21% increase after the second pass with enhanced ductility.

## INTRODUCTION

Ultrafine powders are typically defined by particle size distributions in the submicron range with equivalent diameters of approximately ten to a few hundred nanometers[1]. Ultrafine powders are not novel, but their uses and applications were not realized until engineers and applied scientists realized their superior properties. This was stimulated by the evolution in Nanotechnology through the few past decades. The consolidation behavior of PM products depends mainly on three main parameters, material of the powder, the synthesis method used in producing the powders and the powder consolidation techniques used to produce a bulk product. The main challenge lies in retaining the original properties of the as received metal powders, especially grain size. Conventional approaches used to produce bulk consolidated powders basically include the pressurizing and sintering of powders. The major problem in PM processes is the grain coarsening associated with the hot pressing and sintering processes, which leads to deterioration of the mechanical properties of the component[1,2].

Employing non-conventional techniques for powder consolidation that attracted the research efforts were targeting the retention of the properties of the powder metals, especially

11

when nanostructured materials or nanopowders are employed[3]. Severe Plastic Deformation (SPD) techniques delivered outstanding results in producing Bulk Nanostructured Materials (BNSM) from solid ingot metals and alloys. Although, limited research work was conducted, SPD techniques provided remarkable results in consolidation of metal powder[2-9]. One of the most promising SPD techniques is the Equal channel angular pressing (ECAP). ECAP produces a total equivalent strain of 1.16 mm/mm per pass through simple shear. BNSM produced by ECAP with grain size one order of magnitude less than that produced by conventional forming techniques were reported[2, 3]. Processing consolidated ultrafine structured powders by ECAP is believed to provide a solution for the grain growth associated either with the conventional forming steps. Processing via ECAP should facilitate the use of lower end of the temperatures range used (0.7Tm) during powder compaction stage. This in turn will promote the production of bulk products with superior properties due to the ability to retain the initial nanostructure of the as received powders. Accordingly, ECAP was employed as a secondary processing step following hot compaction of AA2124 nanopowder synthesized via high energy ball milling aiming for the production of BNSM with minimum coarsening of the initial nanocrystalline structure and with enhanced uniform mechanical properties.

## PROCESSING AND CHARACTERIZATION

Gas atomized powder of Al-2124 with a chemical composition of Al-3.9 Cu-1.5 Mg-0.65 Mn-0.1 Si-0.1 was milled for 36 hours in a high energy ball mill at 500 rpm to produce the nanocrystalline nanopowders used in this work. Details for the milling parameters were presented in another publication[10]. The micron powder had a particle size of 40μm and grain size of about 700nm. The ball milled nanopowder particle size was formed of less than 300nm particle clusters (Figure 1a) with about 20nm internal structure in average size (Figure 1b). The powder was first hot compacted using single sided uniaxial pressure in a confined die of diameter 12.7 mm and height 52 mm which produced compacts of a height to diameter (h/d) ratio of 4. Hot compaction of nanopowder took place under combinations of temperatures (360, 420 and 480°C), durations (60, and 90 minutes), and pressures of 450, 525, and 600MPa. All hot compacts were cooled down to 30°C using water quenching. Only intact hot compacts were selected for further processing via ECAP single pass and second pass route (A).

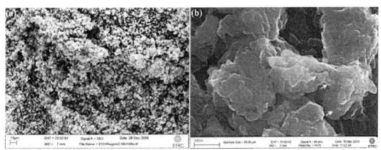

Figure 1. SEM images for the AA2124 nanopowders (a) particle morphology and (b) internal structure

ECAP billets were prepared by inserting the resulting hot compacts in AA2024 cans of 22.8 mm diameter and 80 mm length. The machined cans were over aged for 3 hours at 400°C, and cooled in furnace. Figure 2a shows a schematic for the can with the HC rod inside it. During ECAP, the canned hot compacts were pressed through a die of two equally sized channels 23mm in diameter intersecting at an angle $\phi$ of 90°. Pressing was performed at a ram velocity of 2mm/min and temperature of 245°C on a 500 Tons capacity MTS universal testing machine. The deforming temperature was the lowest temperature that is suitable for producing uniform intense plastic straining via ECAP without causing shear banding or shear localization.

A Leica optical microscope and LEO Supra55 Field Emission Scanning Electron Microscope (FE-SEM) were used for microstructural characterization at the mid plane along the longitudinal direction of the pressed rod. Density of hot compacts and ECA pressed consolidates was measured using the digital Mettler-Toledo densitometer and Xylene as an auxiliary liquid. Density uniformity across the height of the consolidated rod was characterized by measuring the variation in relative density as a function distance away from the compression head for the HC and along the length of the rod for the ECAP ones. The measured slop was a direct measure for the extent of density uniformity produced before and after ECAP. Hardness of the hot compacts and consolidates was measured using the Mitutoyo VMH 810 micro hardness tester. VHN values were measured at 200 gf and 15 seconds dwell time. Hardness measurement and microstructural imaging was conducted on longitudinal surfaces cut parallel to the compaction loading for the HCs and the flow direction for the ECAP rods (Figure 2b). Unidirectional compression testing was conducted on MTS 810 universal testing machine. For each compact, all measurements were taken at the top, middle and bottom sections in the transverse direction from a minimum of three samples.

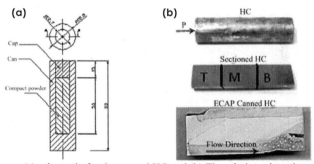

Figure 2. Shows (a) schematic for the canned HC and (b) The whole and sectioned HC before and after ECAP. Top section (T), Middle section (M) and Bottom section (B)

RESULTS AND DISCUSSION

Hot Compaction

Only the compaction conditions (525MPa, 480°C, 90min) and (600MPa, 480°C, 60min) produced intact and crack- free HCs. Table 1 lists the physical and mechanical properties measured for the selected two HCs. The effect of softening by the employed temperature had a significant role in the densification process. Since the synthesis technique employed for the production of the nanopowder particles was milling in a high energy ball mill for 36 hours, the ball milling process introduced internal stresses and caused strain hardening of the powder particles by continuous impact caused by the balls to the powder particles. It is suggested that the strain hardening of the nanopowders resulted in the resistance of the particles to plastic deformation and hence retarded densification; although, the applied compaction pressures were equivalent to 7 and 8 multiples of the AA2124 yield strength.

Table 1. Physical and mechanical properties of selected HCs

| Compaction Conditions | Relative Density (%) | Hardness (VHN) | Compressive strength (MPa) | Grain Coarsening (%) |
|---|---|---|---|---|
| 525MPa, 480°C,90min | 97.4 | 167.7 | 577.1 | 220 |
| 600MPa, 480°C,60min | 95.2 | 181.6 | 537.2 | 190 |

Figure 3 shows OM images for both compaction conditions. Hot compaction at 525MPa for longer duration of 90min produced fine scattered porosity are revealed within the HCs, while higher pressures at shorter durations resulted in the formation of large voids in addition to the fine ones. This agrees with the low measured relative densities of the nanopowder HCs, which resulted in low fracture strains under compressive stress. Voids and cavities, especially the large ones act as stress raisers under loading, which results in crack nucleation, propagation followed by failure. It is suggested a combination of lower pressures (525MPa) and longer durations (90min) at 480°C allowed for the softening of the energetic powder particles boundaries. This in turn facilitated particle sliding within the empty sites and promotes interparticle boundary diffusion with time. Higher densification and consolidation enhanced the overall mechanical properties of the produced compacts. Although longer compaction duration enhanced the compressive strength, it resulted into lower VHN-values, which could be due to recovery associated with heating for a longer time.

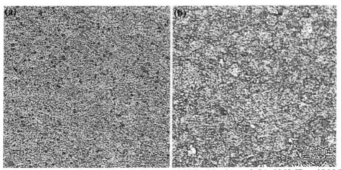

Figure 3 OM micrographs for HC (a) 525MPa, 480°C, 90min and (b) 600MPa, 480°C, 60min.

Post Forming Via ECAP

The macroscopic assessment or visual inspection of the HCs after ECAP is a basic test to investigate whether any cracks or voids were generated during processing. Such cracks appear in the ECAP consolidates either due to opening of inherent defects in the HC or due to shear localization induced by poor lubrication and/or low forming temperature[11,12]. Figure 4 (a-to-d) shows the produced nanopowder consolidate after ECAP one pass and two pass route (A) for the two HCs. Rout A pass represent the condition where the canned HC is fed in the same direction between passes without rotation. After the first pass, the measured angle of inclination of the structure was almost 30° (Figure 4a, c). This disagrees with the theoretical angle (45°)[12]. This implies that the ECA pressed consolidates experienced a lower degree of shear straining compared to the theoretically anticipated values. The observed lower angles of inclination have been explained by Elkhodary et al.,[11] who attributed this behavior to the powder particles rotation when densification is initially incomplete. A similar observation applies to the second pass (A) where the measured angle was 16° versus a 27° theoretical angle as shown in Figure 4 (b and d).

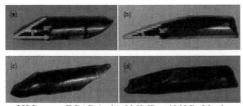

Figure 4. Macrographs of HCs post ECAP (a, b) 525MPa, 480°C, 90min and (c,d) 600MPa, 480°C, 60min 1pass and 2 passes (A) respectively.

Relative Density:

The relative density dropped by about 1% after processing via one pass ECAP, which was followed by a slight increase after 2 Passes route (A) (Figure 5). The decrease in density after the first pass of ECAP can be attributed to the insufficient amount of softening associated with warm forming at 245°C that is needed to overcome the hardening induced by prior milling. The partially consolidated particles by prior HC experienced a partial rigid body rotation at the shear plane instead of being fully strained[11], which agrees with the lower shear angles measured after the first pass deformation (Figure 4 a, c). During the second pass, the HCs had already been subjected to an additional heating cycle at 245°C. This could have resulted in strain softening of the remaining unconsolidated hardened powder particles and hence enhanced the effectiveness of ECAP[3], which explains the observed increase in density during the second pass.

Density Uniformity:

Table 2 shows the average slopes of the density plots of the HCs before and after ECAP one pass and 2 pass route (A). Comparing the slopes of density plots produced for the nanopowder HCs before and after ECAP reveals the enhancement in density uniformity across the produced rod length after the first pass, while the density uniformity was doubled after the second pass route (A). It is observed that the higher the degree of uniformity in density produced by prior hot compaction, the higher the enhancement after processing via ECAP.

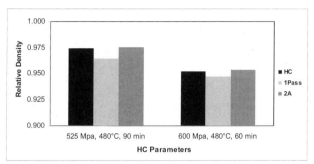

Figure 5. Relative density variation for the HCs before and after ECAP one Pass and 2 Pass (A)

Table 2. Average slopes for HCs before and after ECAP one and 2-pass route (A)

| Compaction Conditions | HC slope | ECAP | | Improvement (%) |
|---|---|---|---|---|
| | | Processing route | Average slope | |
| 525MPa, 480°C,90min | 0.01 | 1-Pass | 0.0028 | 357 |
| | | 2-Pass (A) | 0.0013 | 769 |
| 600MPa, 480°C,60min | 0.012 | 1-Pass | 0.005 | 240 |
| | | 2-Pass (A) | 0.0025 | 480 |

Microhardness

Hardness of the HCs before and after the first and second pass (A), is shown in Figure 6. In general, processing via ECAP one pass resulted in 5% decrease in hardness compared to that of the HCs. After the second pass the hardness experienced an additional 8% decrease compared to that of the first pass consolidates. The drop in hardness after each ECAP pass reflected the dominant effect of strain softening over strain hardening that is induced by heating during ECAP. The energetic condition of the milled powder particles should result in a higher tendency of the structure to dynamic recovery during ECAP.

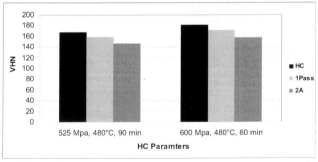

Figure 6. Hardness variations for the HCs before and after ECAP one Pass and 2 Pass (A)

The 600MPa, 480°C, 60min consolidates maintained higher VHN-values than that of the corresponding 525MPa, 480°C, 90min ones. Microhardness is a measure of the mechanical properties on the micron scale, which represents the degree of strain hardening and bond strength of the consolidated powder particles as a function of the processing stages[13]. Accordingly, retention of the hardening induced by prior HC conditions influenced directly on the produced hardness by subsequent ECAP passes. The produced results for hardness agree with the findings of Karaman et al.[14] and Murayama et al,[15] who reported hardness decrease after multiple passes of pure Cu powder due to the susceptibility of grain growth when experimenting copper powder green compacts extruded at room temperature.

Compressive Strength:

A representing behavior of the nanopowder HCs under uniaxial compression loading before and after ECAP is shown in Figure 7. The HCs displayed a relatively low amount of plastic deformation prior to failure and high compressive yield and ultimate strength (Figure 7). This behavior resulted due to the relatively low density of the consolidates and the nature of the strain hardened powder particles that could hardly experience plastic deformation under compression. Similar to hardness, the HCs compressive behavior controlled that produced by subsequent one and 2-pass ECAP.

The compressive strength of the HC was highest followed by 2-pass route (A) processed rods and lowest displayed after the first pass. This agrees with the recorded decrease in relative density after the first pass followed by an increase after the second pass (Figure 5). Since compression testing represents a measure of the overall mechanical properties on the macro-scale, increasing porosity content affects directly the compressive behavior of the HCs. This agrees with the results reported by Bing et al., who proved that the compressive properties of nanostructured Al-alloys consolidates are directly influenced by their degree of densification[16].

Figure 8 shows a chart diagram that compares compressive strength of the nanopowder HCs before and after ECAP. The 525MPa, 480°C, 90min consolidates maintained higher values of compressive strength over the 600MPa, 480°C, 60 min consolidates after the first and second pass. This reflects that the input HC compressive strength controlled that of the produced consolidates after subsequent ECAP.

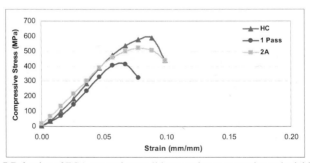

Figure 7 Behavior of ECA pressed consolidates under compressive uniaxial loading

For both compaction conditions, the compressive strength decreased after the first pass by about 50% then it increased after the second pass by about 26% resulting in lower compressive properties compared to the HC. The observed drop in compressive properties after the first pass could be attributed to the measured decrease in density of HCs during pressing (Figure 5); in addition to the strain softening associated with the warm deformations. The relative increase in compressive strength after the second pass also followed the increase in density of the consolidates (Figure 5), which was enough to increase the effectiveness of the strain hardening induced by ECAP. Moreover, higher densify of the HCs prior to ECAP increases the effectiveness of the process through the increased utilization of a total strain per pass of 1.16mm/mm [16].

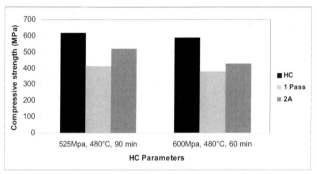

Figure 8. Compressive strength variations for the HCs before and after ECAP one and 2 Pass (A)

Microstructural Evolution:

In order to explain the influence of ECAP one and 2-pass processing on the mechanical behavior of the nanopowder HCs, it was necessary to investigate the microstructural evolution as a function of the various processing stages. Figure 9 shows OM images for 525MPa, 480°C, 90min compared to the 600MPa, 480°C, 60min HCs after subsequent one and 2-pass ECAP. Grains could hardly be revealed due to high degree of deformation for the nanopowder HC after ECAP (Figure 9 a-d). From the images, extent of densification was manifested in the form of voids, and cavities that were scattered within the matrix. It is clear that high porosity content in the nanopowder HCs were present even after the second pass.

The displayed images at low magnification also revealed the inclination angles of the developed structures, which varied based on the compaction parameters and ECAP number of passes[11]. Consolidates of 525MPa, 480°C, 90min exhibited higher angles of 32° and 18° compared to 28° and 16° for those processed at 600MPa, 480°C, 60min (28° and 16°) after the first and second passes via ECAP, respectively. This was attributed to the higher density of the 525MPa, 480°C, 90min HCs. This indicates that the lower initial density input to ECAP the higher the energy loss (lower total strain per pass) by rigid particle rotation within the shear zone.

Figure 10 shows SEM images for both HC conditions before and after one and 2-pass ECAP. Equiaxed voids were revealed in both HC conditions as shown in Figure 10 (a and d) for the 525MPa, 480°C, 90min and 600MPa, 480°C, 60min hot compacts, respectively. Images for the higher pressures compacts revealed an evidence for crack nucleation and propagation at the coarse second phase particles (Figure 10d), while an evidence for unconsolidated nanoparticles clusters was manifested for both HC conditions. It is suggested that compaction pressures of 600MPa over short durations of 60min resulted in inducing relatively high residual stresses of the strained powders around the coarse second phase particles, which could have resulted into crack nucleation either during compaction or during the subsequent water quenching. Moreover, ECAP processing one-pass was unable to eliminat the unconsolidated clusters for the high pressure and short duration HCs (Figure 10e). Conversely, one pass ECAP images of the lower pressure and longer duration HCs showed no evidence for such clusters (Figure 10b). Elongated voids in the direction of shear are also observed after one pass for both conditions. ECAP 2-pass via route A, show almost no voids whether elongated or equiaxed within the deformed structure of the 525MPa, 480°C, 90min compacts (Figure 10c), while a significant fraction of sheared voids are still present within the structure of the 600MPa, 480°C, 60min ones (Figure 10f). Combination of crack nucleation, high porosity and unconsolidated powder clusters within the consolidated matrix promoted the lower compressive strength and lower ductility of the 600MPa, 480°C, 60min compacts (Figure 8), which negatively influenced the one pass ECAP produced consolidates.

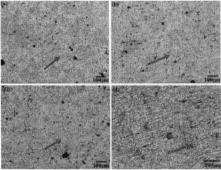

Figure 9 OM images for the HCs post ECAP one-pass and 2-pass (a,b) 525MPa,480°C,90min and (c,d) 600MPa, 480°C, 60min, respectively

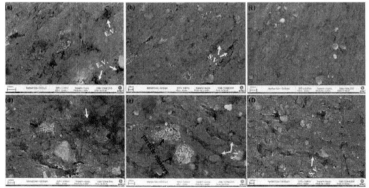

Figure 10 SEM images of HCs before and after one and 2-pass ECAP (a-c) 525MPa, 480°C, 90min and (d-f) 600MPa, 480°C, 60min, respectively. Equiaxed voids (E.S.) Sheared voids (S.V)

HRSEM and TEM imaging employed on thinned and electropolished samples was employed in an attempt to reveal the structure of the nanopowder consolidates after ECAP one pass. Figure 11a shows HRSEM image for the 600MPa, 480°C, 60min consolidate after one pass. The image shows a heavily deformed structure, oriented in the direction of shear with no details of the size of the structure. TEM imaging revealed the formation of structure ~200nm in size with evidence of high dislocation activities on the inside and at their boundaries in the form of extinction contours. Other regions showed evidence for finer structure about 54nm in size formed within the coarser ones (pointed at by an arrow) with clear boundaries as shown in Figure 11b. This agrees with the grain size measured using SEM images shown in Figure 10. Second phase particles with different morphologies are shown within the SEM (Figure 10) and TEM (Figure 11b) images. It is well known that AA2124 is a complicated heat treatable alloy due to the formation of a variety of second phase particles and precipitates, which strongly depends on the thermomechanical treatment envolved[17]. Effect of the various stages of processing on the morphology of the phases present especially in the T851 condition is currently under investigation.

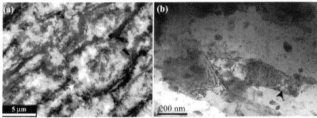

Figure 11. Images for the nanopowder HC 600MPa, 480°C, 60 min consolidate after 1 pass ECAP (a) HRSEM, (b) TEM

Figure 12 shows the average grain size measured for each of the nanopowder HCs before and after ECAP. It is very important to note that the produced consolidates after hot compaction at 480°C followed by two passes ECAP at 245°C still retained a nano sized structure <62 nm. The 525MPa, 480°C, 90min consolidates maintained a coarser average grain size compared to the 600MPa, 480°C, 60 min consolidates, which was attributed to the initial coarsening that took place during the HC stage. It is shown that grain size increased with increasing the number of passes, which agrees with Bang *et. al.*[16].

Table 3 lists the calculated % grain coarsening encountered for both compaction conditions as a function of the processing stages compared to the initial as-milled powder grain size (20nm). It is obvious that most of the grain coarsening took place during the HC stage followed by a slight coarsening via ECAP passes; especially after one pass. Although, an average total coarsening of around 280% occurred via the combined processing of the heavily strain hardened nanopowders HC/ECAP, the enhancement in relative density, density uniformity and mechanical properties indicates the importance employing both techniques. From Figure 7, ECAP 2-pass rout (A) produced a combination of high compressive strength with relatively enhanced ductility compared to the first pass. Based on the displayed results, it is anticipated that increasing the number passes beyond 2-passes could improve the consolidation and hence enhance the product physical; and mechanical properties. Comparing the AA2124-T851 properties with that produced for the investigated 36hr milled nanopowder consolidates properties without tempering heat treatment, it is found that a hardness of 146 versus 181HV were produced, respectively[17]. In addition, the processed nanopowders display compressive strength of 590-620MPa, which is equivalent to metal matrix composite of AA2124 micronpowder (75μm particle size) reinforced with either 10% BN or 15% $Al_2O_3$ ceramic powders processed by a combined cold uniaxial compaction and conventional extrusion at 480°C with an extrusion ratio of 3.25:1[18].

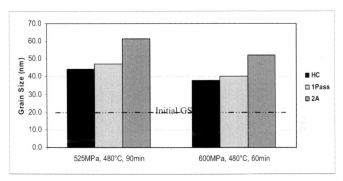

Figure 11. Grain size variations for the HCs before and after ECAP one Pass and 2 Pass (A)

Table 3 % grain coarsening of the initial nanopowder post processing via hot compaction followed by one and 2-pass ECAP

| Compaction conditions | | | ECAP | HC grain coarsening (%) | ECAE grain coarsening (%) | Total Grain coarsening (%) |
|---|---|---|---|---|---|---|
| Pressure (MPa) | Temperature (°C) | Time (min) | | | | |
| 525 | 480 | 90 | 1 Pass | 220 | 15 | 235 |
| | | | 2 Pass (A) | | 87 | 307 |
| 600 | 480 | 60 | 1 Pass | 190 | 10 | 200 |
| | | | 2 Pass (A) | | 71 | 261 |

CONCLUSIONS

1. AA2124 nanopowders fabricated by high energy ball milling with an initial grain size of 20nm were consolidated into bulk nanostructured rods 50-60nm in grain size by a combined processing of HC and ECAP with uniform superior mechanical properties.
2. Parameters of the HC stage controls and influences significantly the densification degree and hence the mechanical properties produced after subsequent ECAP passes.
3. Although, processing one pass ECAP enhanced the consolidates density uniformity, it resulted in lower density, hardness and compressive strength compared to those of the HC due to rigid body rotation of the partially unconsolidated particle clusters inherited from prior stage of HC.

4. ECAP 2-pass route (A) improves compressive strength and ductility and provides an insight for improved compressive properties when multiple passes are employed.
5. The processed bulk products without post heat treatment demonstrate properties higher than the AA2124-T851 alloy and equivalent to dispersion strengthened AA2124 metal matrix composite reinforced with 10-15% $Al_2O_3$.

ACKNOWLEDGMENTS:
Authors of the work would like to acknowledge the financial support of the Yousef Jameel Science and Technology Research Center (YJ-STRC) at the American University in Cairo for the financial support of the current research work. Moreover, we would like to acknowledge Dr. M. Attallah from Manchester University for conducting the TEM and HRSEM investigation.

REFERENCES:
1. M. Galanty, P. Kazanowski, P. Kansuwan, W. Misiolek: J. Mat. Proc. Tech., (2002) pp.491-496.
2. Senkov, D.B. Miracle, J.M. Scott, S.V. Senkova, Materials Science and Engineering A, (2003), pp.12–21.
3. R. Valiev, T. Langdon: Progress in Materials Science, (2006), pp. 881–981.
4. N. Senkov, S. V. Senkova, J. M. Scott, D. B. Miracle: J. Alloys Compd., 2004, vol. 365, pp. 126-33.
5. J. M. Scott, Kending, O. N. Senkov, D. B. Miracle, Watson, Davis, An Odyssey of Materials in Space, in: TMS Annual Meeting, New Orleans, LO, USA, 2001.
6. S. Xiang, K. Matsuki, N. Takatsuji, M. Tokizawa, T. Yokote, J. Kusui, K. Yokoe: J. Mater. Sci. Lett., 1997, vol. 16, pp. 1725-27.
7. N. Senkov, S. V. Senkova, J. M. Scott, D. B. Miracle: Mater. Sci. Eng., A, 2005, vol. 393, pp. 12-21.
8. K. Xia, X. Wu: Scripta Mater., 2005, vol. 53, pp. 1225-29.
9. J. Robertson, I. Karaman, H. K. Hartwig, J. Anderson: J. Non-Cryst. Solids, 2003, vol. 317, pp. 144-51.
10. H. Salem, S. El-Eskandarany and H. Abdul Fattah: "Characterization of the Consolidation behavior of Fabricated Nanocrystalline-Nanopowders of TiC/Al-2124 Composite", ASME Multifunctional Nanocomposite conference and Exhibition, September 20-22, (2006), Honolulu, Hawaii.
11. K. I. Khodary, H. G. Salem and M. A. Zikry: Metallurgical and Materials Transactions A, Accepted March 8th 2008. in press
12. T. Langdon, "Journal of Materials Science and Engineering, (2007), pp. 3–11, 2007.
13. T.S. Srivatsan, B.G. Ravi, A.S. Naruka, M. Petraroli, R. Kalyanaraman, and T.S. Sudarshan: Materials and Design 2002, pp. 291_296J.
14. Karaman, J. Im, S. Mathaudhu, Z. Luo and K. Hartwig: Metallurgical and Materials Transactions A , Vol. 34A, (2003), pp. 247-256.
15. M. Murayama, Z. Horita and K. Hono: Acta Materialia, Vol.49 (2001), pp. 21-29
16. Q. Bing, Q. Han and J. Lavernia: Adv. Eng. Mat., Vol. 7 No 5, (2005) pp. 457-467.
17. L. Dobrzanski, A. Wlodarczky and M. Adamiak: J. Mat. Pro. Tech., Vol. 175 (2006) pp.186-191.
18. ASM Materials Data Sheet ASM Aerospace Specifications Metals "Aluminum 2124-T851", http://asm.matweb.com/search/specificmaterials.asp, 7/14/2008.

# LARGE-SCALE (>1GM) SYNTHESIS OF SINGLE GRAIN TWO-PHASE $BaTiO_3$-$Mn_{0.5}Zn_{0.5}Fe_2O_4$ NANO-COMPOSITES WITH CONTROLLED SHAPES

Yaodong Yang,* Shashank Priya, Jie-Fang Li and D. Viehland
Department of Materials Science and Engineering, Virginia Tech
Blacksburg, VA 24061 USA

A BSTRACT
The development of a gram-scale synthesis method with reproducible particle geometries is a key to the realization of nano-materials. Here, we have developed such a large-scale synthesis of single grain, two-phase $BaTiO_3$-$Mn_{0.5}Zn_{0.5}Fe_2O_4$ nano-particles with controlled shapes, offering the possibilities of a new class of integrated multi-ferric materials.

## INTRODUCATION

Multi-functionality in composites requires the bringing together of two or more materials with dissimilar structures. This is done in order to achieve the optimization of two or more independent properties: for example, magnetization and polarization in magnetostrictive / piezoelectric composites.[1,2]

Composites of dissimilar functionalities can have unique product tensor properties, which neither phase possesses individually[3]: for example, magnetoelectricity. Magnetoelectric or ME composites of various length scales have been reported to be fabricated by various physical and chemical methods. The simplest case being laminated composites of magnetostrictive alloys (Terfenol-D, Meglas, Galfenol) and piezoelectric oxides ($Pb(Zr_{1-x}Ti_x)O_3$ or PZT, $Pb(Mg_{1/3}Nb_{2/3})O_3$-$PbTiO_3$ or PMN-PT) that are of mm or larger size.[4]

Another, more complex, method by which ME composites have been fabricated is pulsed laser (PLD)[5] or physical vapor (PVD) deposition.[6] Epitaxial thin film nano-structured composites of ferromagnetic and piezoelectric oxides have been reported with various geometric arrangements of phases. These include (i) self-assembled two-phase nanocomposites consisting of magnetostrictive $CoFe_2O_4$ (CFO) nanorods in a $BaTiO_3$ (BTO) matrix grown on $SrTiO_3$ (STO) substrates : a single layer approach;[6] and (ii) epitaxial heterostructures consisting of CFO thin-layers grown on PZT ones, which were previously grown on STO substrates: a layer-by-layer approach with a sandwich structure.[7] However, vapor deposition methods of epitaxial composites have two obvious disadvantages: high cost, and limited quantity of yields.

A low cost method, Sol-gel, has been used to prepare metal oxide nanostructures.[8] Xie et al. have made magnetostrictive ferrite / perovskite ferroelectric ME nano-composites, even as nanowires, via electrospinning.[9] A concern with these proceses is the use of organometallic precursors which require sophisticated handling and will present environmental concerns. Alternatively, one could prepare two-phase (ferrite-perovskite) nanoparticles by solid state reactions. However, such self-assembled 'crystals' prepared by an environmentally-friendly solid state method has yet to be reported. This would be interesting because the exchange between magnetostrictive and piezoelectric phases is mediated via their strictions: i.e., changes in the geometrical shape of particles could impact ME properties.

## EXPERIMENT

In this study, we demonstrate a large-scale and easily controllable solid-state reaction method to prepare two phases BaTiO$_3$-Mn$_{0.5}$Zn$_{0.5}$Fe$_2$O$_4$ (BTO-MZF) in one nano-particle with controllable shape and size. Barium acetate, TiO$_2$, MnO, ZnO, Fe$_2$O$_3$, NaCl and NP-30 (nonylphenyl ether) were mixed with corresponding ratios (4:4:1:1:4:120:20 for BTO rich samples, and 1:1:1:1:4:120:20 for MZF rich ones), grinded (25 min), and sonicated (10 min) to make the mixture uniform. NP-30 was used as nonionic surfactant to help with uniform mixing. The mixture was then annealed at 850°C for 5 hrs. After cooling to room temperature, the samples were washed with distilled (DI) water and dried in an oven overnight at 80°C. The final compositions of the two-phase material that were synthesized correspond to: (BaTiO$_3$)$_2$-Mn$_{0.5}$Zn$_{0.5}$Fe$_2$O$_4$ (designated here as B2M1) and BaTiO$_3$- (Mn$_{0.5}$Zn$_{0.5}$Fe$_2$O$_4$)$_2$ (designated here as B1M2). The special shape of the particles was achieved by controlling the sold-state reaction temperature and time.

The two phase equilibria of the particles was first confirmed to be spinel-perovskite by x-ray diffraction using a Philips MPD system. Next, the composition was determined using a PHI Quantera SXM scanning photo electron spectrometer. Finally, the morphologies of the particles was determined using a LEO (Zeiss) 1550 scanning electron microscope (SEM).

## RESULT AND DISCUSSION

In Figure 1, we present a XRD line scan for B2M1 as an example. All of the peaks in these figures could be indexed to either perovskite BaTiO$_3$ or spinel Mn$_{0.5}$Zn$_{0.5}$Fe$_2$O$_4$, clearly demonstrating the presence of a two-phase equilibria. Please note that the relative intensity of the peaks changed with volume fractions of the two phases, without the appearance of new peaks indicative of additional phases. As can be expected, for B2M1 the strongest peak belonged to MZF. From the Fig.1 we can see the highest peak belonged MZF (311) and a peak around 56° was related to MZF (511) and BTO (211) that was the reason why it was broader than other peaks. The MZF structure was cubic for the B2M1 samples. And the BTO was found to exhibit a perovskite structure. One can probably explain this by taking into account the shape of the particles. For B1M2, the shape of the particles was pyramidal which has resemblance to the spinel structure and whose surface might be MZF rich. For B2M1, the shape of particles was rod, which would make it difficult for the MZF to preferentially occupy the surface.

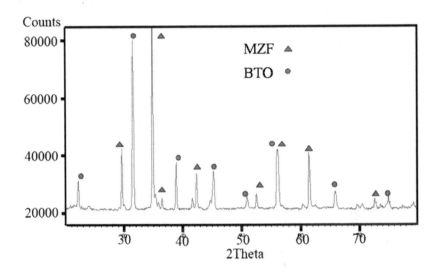

Figure 1. XRD spectrum of B1M2 spinel sample.

The atomic ratios were then determined by x-ray photoelectron spectroscopy or XPS. The results are summarized in Table I. Inspection of this table will show that the elemental ratios of Ti:Fe were 2.8:5.9 for B1M2 and 5.9:2.9 for B2M1. These results confirm that the two phases composition of our samples is quite close to the cation stoichiometry that we had proportioned during solution processing. However, the concentrations of Mn and Zn were less than the initial stoichiometry, possibly because of poor dispersion or loss during firing. The minor changes in chemistry could also be related to defects at the surface, which often form strain-minimized surfaces.

Table I. Atomic ratio from two samples

| Atomic | O1s | C1s | Ti2p | Ba3d5 | Cl2p | Fe2p3 | Mn2p3 | Zn2p3 |
|--------|-----|-----|------|-------|------|-------|-------|-------|
| B1M2 | 48.4 | 37.0 | 2.8 | 3.3 | 1.3 | 5.9 | 0.7 | 0.6 |
| B2M1 | 69.9 | 10.2 | 5.8 | 5.2 | 4.9 | 2.9 | 0.6 | 0.6 |

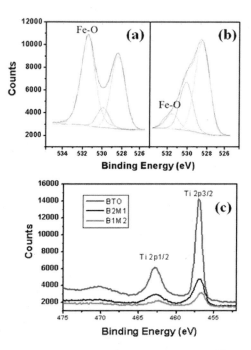

Figure 2. XPS spectrums of O1s on (a) B1M2 spinel particles (b) B2M1 nanorods and (c) Ti2p scan on three samples. Here BTO (BaTiO3) used as control sample.

Figure 2 shows XPS data for (a) B2M1, and (b) B1M2. By comparing these three XPS spectra, we can see that the intensity of the Fe-O bond at 531.4eV increased notably with increasing concentration of MZF: for pure BTO, no such peak was found (data not shown). Please note that the peak positions slightly shifted towards higher binding energies for the rod type structure. Because rods (1-dimension) require more surface area than particles (0-dimension), the coordination is expected to be different for these two cases. We next obtained XPS spectra of Ti for BTO, B2M1 and B1M2, as shown above (Fig. 2c). All three samples yielded similar spectra. There were no additional peaks found, and the bond energy of Ti 2p$_{1/2}$ and 2p$_{3/2}$ shifted to only slightly lower energy with increasing Fe content. These results show that Ti has nearly the same chemical environment in the BaTiO$_3$ phase of all three samples, and that the Ti-O bond is only slightly weakened by being placed in an environment rich in a second phase containing a high concentration of Fe-O bonds.

Figure 3 shows SEM images of the grain morphology for (a) pure MZF and (b) pure BTO. Pure MZF can be seen to have pyramid shaped grains with a size of about 100nm. The structure is a cubic spinel (Fd3m)[10]. Pure BTO can be seen to have a smooth surface and hemispherical tips. The particle size was less than 1μm in width and about 5μm in length.

Figure 3. SEM image of (a) pure MZF spinel nanoparticles, (b) pure BTO nanorod, (c) B1M2 spinel nanoparticles, and (d) B2M1 nanorods. (e) Fe element map. Insert is SEM image of element analysis focus area. (f)Ti element map. All of these three images are a same particle whose size is around 1μm.

We then determined the grain morphologies for B1M2 and B2M1, as shown in Figures 3c and 3d, respectively. For B1M2, the grain geometry had eight faces in each hemisphere, whereas that of pure MZF had only four. A similar change was found in B2M1. Compared with pure BTO nanorods that have more surface area, B2M1 rods had eight faces too. We also found that the particle size could be readily controlled by adjusting the solid state reaction time (Fig.4).

Figure 4. SEM images of following samples: (a) B1M2 spinel nanoparticles prepared by reaction for 1h, (b) high magnification image of a, (c) 5h, (d) 7.5h.

Next, we performed elemental mapping of Fe and Ti, in order to prove that the B1M2 and B2M1 particles contained both Ti and Fe. Figures 3e and 3f show the Fe- and Ti-contrasts for B1M2, respectively. Both elements can be seen to be well-distributed across the entire particle. Similar results were found for B2M1 rods (data not shown here). These results establish that individual B1M2 and B2M1 particles consist of both spinel and perovskite phases, distributed there within.

CONCLUSION

Our results demonstrate that we can synthesize single particle, two-phase MZF-BTO nano-composites by solid state reaction. We find a cross-over with changing volume fractions of the two phases between particles from shapes resembling those of pure MZF to those of pure BTO. The extra faces formed in our grains is interesting, and may result from the two phases having different lattice parameters, where there is a need to minimize the internal strain in addition to surface energies. Pure MZF is a cubic spinel (Fd3m) with ferrimagnetic properties, whereas pure BTO is a classic ferroelectric perovskite: integrated into a single particle, it offers the possibility to create a new class of multiferroic materials.

Presently, we are working on synthetic optimization and the fabrication of simple ME device based on our two-phase nanomaterial. Because these nanoparticles/rods are magnetic we can control particle distributions by engineering patterned microstructure via external magnetic field. Various patterns on different substrates could easily be fabricated by this process. It offers potential device designs for integrated microelectronics.

ACKNOWLEDGMENT

Support of this work was provided by National Science Foundation. We also thank NCFL in VT for SEM work and Mr. Makarand Karmarkar for his kind help.

FOOTNOTES

* Author whom correspondence should be addressed E-mail: yaodongy@vt.edu

REFERENCES

[1] J. Ryu, S. Priya, A. V. Carazo, K. Uchino and H. E. Kim, J. Am. Chem. Soc., **84**, 2905, (2001)

[2] V. M. Petrov, G. Srinivasan, M. I. Bichurin and A. Gupta, Phys. Rev. B, **75**, 224407, (2007)

[3] W. Carlson, D. Williams and R. Newnham, J. Electroceram., **2**, 257, (1998)

4 See, for example: a) S. X. Dong, J. Y. Zhai, J. F. Li and D. Viehland, Appl. Phys. Lett., **89**, 252904, (2006) b) S. X. Dong, J. Y. Zhai, J. F. Li and D. Viehland, Appl. Phys. Lett., **89**, 122903, (2006)

[5] H. Zheng, F. Straub, Q. Zhan, P. Yang, W. Hsieh, F. Zavaliche, Y. Chu, U. Dahmen and R. Ramesh, Adv. Mater., **48**, 2747, ( 2006)

[6] Z. Lu, H. Cheng, M. Lo and C. Y. Chung, Adv. Funct. Mater., **17**, 3885, (2007)

[7] C. Y. Deng, Y. Zhang, J. Ma, Y. H. Lin and C. W. Nan, J. Appl. Phys., **102**, 074114. (2007)

[8] Y. Yang, L. Qu, L. Dai, T. S. Kang and M. Durstock, Adv. Mater., **19**, 1239.( 2007)

[9] S. H. Xie, J. Y. Li, Y. Qiao, Y. Y. Liu, L. N. Lan, Y. C. Zhou and S. T. Tan, Appl. Phys. Lett., **92**, 062901, (2008)

[10] Y. Mao, S. Banerjee and S. S. Wong, J. Am. Chem. Soc., **125**, 15718, (2003).

# PROPERTIES OF ALUMINA DIELECTRICS VIA INK JET PROCESS

Eunhae Koo[*], Yoo-Hwan Son, Hyunwoo Jang, Hyotae Kim, YoungjoonYoon, Jong Hee Kim

Division of Fusion and Convergence Technology,
Korea Institute of Ceramic Engineering & Technology, Seoul 153-801, Korea.
*Email: ehkoo@kicet.re.kr

## ABSTRACT

Low loss dielectrics of ~1 μm have been fabricated on the Si wafer via ink jet process using alumina dielectric inks formulated with alumina powders and anionic polymer dispersants in formamide/water. The ink chemistry and rheological properties are important in printing the smooth layer without clogging the nozzle orifice. The inverse of the Ohnesorgi's number ($Z^{-1}$) can provide useful criteria for jettable ink. The Marangoni flow induced by appropriate mixed solvent helps the dielectric to be flat layer suitable to high density interconnects with embedded passives. The volume fraction of alumina of the dielectric layer by ink-jet process is about 70% much higher than that of the dielectric of about 40% fabricated by a conventional wet casting method. Furthermore, the Q factor of the dielectric was evaluated by the impedance analyzer. The results suggest that the dielectrics with high Q factor and dimensional stability can be used for high speed electronic applications

## INTRODUCTION

Direct writing process has many advantages in fabricating electronics over conventional vacuum process based on photolithography/etching process. It can be a very cost-effective process with high productivity. In addition, it is an environmentally benign process not wasting materials because it just patterns on the demanded position. Recently, there has been many efforts on the ink jet process for developing PCB (printed circuit boards) with high density interconnects and embedded passives, LCD color filter, transistors, micro lenses using ink jet process.[1-6] For high speed and wireless RF communications, low loss materials should be used to transfer the signal without noise. Ceramic materials have been used widely in applications such as SAW filter, FEM module for wireless communication, because most ceramic materials have excellent electrical characteristics than organic based materials.[7] But it has intrinsic weaknesses such as its difficulty to process and brittleness. To overcome these disadvantages, we have developed a new noble strategy to fabricate low loss substrates to be used in three dimensional high density interconnects with embedded passives via ink jet process.

In the present study, we focused on the formulation of ceramic ink suitable to ink jet process to develop low loss dielectric layer for being used in high speed and RF applications. Also, we investigate the microstructure, physical and electrical properties of ceramic dielectrics fabricated by ink jet process.

EXPERIMENTAL PROCEDURE

Alumina inks were prepared using mixed solvent of 75 vol% water and 25 vol% formamide (FA) as a drying control agent. The sizes of alumina powders are 0.2μm (ASFP-20 obtained from Denka), 0.3μm (AKP-30 obtained from Sumitomo), and 0.54μm (A161-SG obtained from Showadenko), respectively. In order to make stable inks, 10 vol% of the powder and dispersant (BYK-111 obtained from BYK Chemicals) was milled in mixed solvent for 12 hrs. The rheological properties of inks were characterized using a cone and plate viscometer (DV-II+, Brookfield Engineering) and a surface tentiometer (Surface Tensiomet 21, Fisher Scientific).

The alumina ink was jetted on a p-silicon wafer of [100] orientation cleaned using pirana solution and then rinsed with acetone and ethanol. The printing of alumina ink were carried out using UJ 200 (manufactured by Unijet Inc.) with 50μm orifice diameter of a piezoelectric nozzle fabricated by Microfab technologies, Inc. The printer head was mounted onto a computer-controlled three-axis gantry system capable of the movement accuracy of ±5μm. A CCD camera with strobe LED light was used to characterize the size and shape of individual droplet. The volume of expelled droplets was 150-160pL, traveling with the velocity of 2.5-3.2m/s. The diameter of jetted droplets was 61-65μm. The microstructure of alumina dielectrics on Si wafer was characterized using a field emission scanning electron microscope (model: JSM6700F from JEOL). The electrical properties were measured using an impedance analyzer (model: HP4194A from Agilent Technology).

RESULTS AND DISCUSSION

Fabricating electronic devices by the ink jet process, the ink should meet the general requirements[8-9] about physical and rheological properties as shown in Figure 1. Moreover, there must be no ink settlement inducing the nozzle clogging, and the inverse of the Ohnesorge's Number ($Z^{-1} = N_{Re}/\sqrt{N_{We}}$, $N_{RE} = \rho \upsilon \delta / \eta$, $N_{We} = \rho \upsilon \delta^2 / \gamma$) should be in the range of 1 to 10. If $Z^{-1}$ is smaller than 1, large pressure is required to pull ink out from the orifice of the nozzle because of the viscous characteristic of ink. However, ink with $Z^{-1}$ above 10 forms a very large liquid column from the nozzle which causes the formation of satellite drops. Figure 2 shows the viscosity and stability (the inset of Figure 2) of alumina ink as a factor depending on the size of alumina powder as well as the volume % of the dispersant.

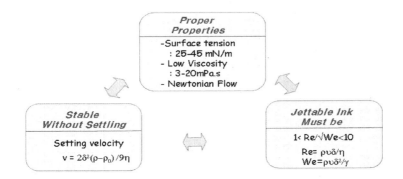

*Figure 1 The general requirements of inks for the microfabrication ink jet process*

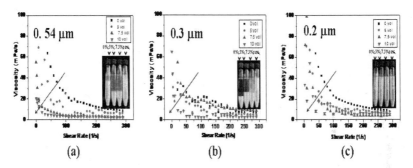

*Figure 2 The viscosity and stability of alumina inks as a factor of the size of alumina powders and the vol% of the dispersant in the ink*

The viscosity of alumina ink made of 0.54μm alumina powder is lower than those of 0.2μm and 0.3μm alumina powder. It stems from the fact that the total surface area of alumina powders in ink suspension decreases as the size of alumina powder increases. This is because the stability is significantly affected by the settling velocity of powders in the ink. Based on the insets of Fig 2, it should be noted that alumina inks of 0.2-0.54μm alumina powders stabilized above 10 vol% BYK can be suitable to the ink jet process without the nozzle being clogged. Based on the Table 1 and Figure 1, it suggest that the alumina inks of 0.2μm alumina with 5-10vol% of the dispersant and 0.3μm alumina with 5vol% of the dispersant are suitable for the ink jet process.

Table 1 The physical and rheological properties of alumina inks

| BYK φ % | ρ (kg m⁻³) | η (m Pa s) | γ mN/m | $N_{Re}$ | $N_{We}$ | $Z^{-1}$ | K | ξ |
|---|---|---|---|---|---|---|---|---|
| 10% vol Al₂O₃ suspension, 0.54 µm | | | | | | | | |
| 0 | 1368 | 22.1 | 62.0 | 9.32 | 7.49 | 3.40 | 1.56 | 1.23 |
| 5 | 1352 | 11.3 | 43.2 | 18.43 | 10.63 | 5.66 | 1.57 | 1.32 |
| 10 | 1215 | 6.4 | 46.1 | 26.33 | 8.94 | 8.81 | 1.32 | 1.45 |
| 10% vol Al₂O₃ suspension, 0.50 µm | | | | | | | | |
| 0 | 1237 | 17.2 | 62.0 | 9.32 | 6.77 | 3.60 | 1.49 | 1.25 |
| 5 | 1218 | 6.9 | 43.2 | 26.08 | 9.40 | 8.49 | 1.36 | 1.43 |
| 10 | 1200 | 4.6 | 46.1 | 33.90 | 8.93 | 11.32 | 1.24 | 1.51 |
| 10% vol Al₂O₃ suspension, 0.20 µm | | | | | | | | |
| 0 | 1336 | 31.6 | 62.0 | 6.34 | 7.31 | 2.34 | 1.70 | 1.14 |
| 5 | 1386 | 16.1 | 43.2 | 12.91 | 10.89 | 3.91 | 1.74 | 1.23 |
| 10 | 1375 | 10.0 | 46.1 | 17.88 | 10.12 | 5.63 | 1.54 | 1.32 |

Figure 3 is the microstructure of alumina dielectrics investigated by FESEM, which shows a very flat and smooth layer of ~1 µm. The smoothness of the dielectrics is important in fabricating the electronic device as designed, in which the noise is caused from the impedance mismatch by the thickness variation of dielectrics. One of the disadvantages in the ink jet process stems from the sticking of solute at the edge of droplets, which causes coffee-ring phenomenon during the drying process.[10-11] To overcome the coffee ring effect, the appropriate solvent system, water as the main solvent (boiling point: 100°C, γ= 72.8 Nm/m @20°C) and formamide as co-solvent (boiling point: 210°C, γ= 58.2 Nm/m @20°C) was selected, in which the inward convective flow is induced due to Marangoni flow.[11-12] When printed layers are dried in the ink jet process, the variation of the ratio of the main solvent to co-solvent in ink-jetted droplets has the gradient of surface tension across the surface of ink-jetted droplets. Park et al[13] reported that Marangoni flow is easily induced when a co-solvent as a humectant has higher boiling point and lower surface tension than those of the main solvent. Calculating from the volume and weight of printed alumina dielectrics, the volume fraction of alumina of the dielectric layer by ink-jet process is about 70% higher than that of the dielectric of around 40% fabricated by a conventional wet casting method.

Figure 4 and 5 show the microstructure and Q-factors of alumina dielectrics infiltrated by polymer resins. As shown in Figure 4, the resin is well infiltrated by the ink jet process without any voids in the dielectric layer. When infiltrating cyanate ester resin, Figure 5 shows that the dielectric with the high Q factor of 200 can be fabricated by the sequential ink jet process. These results suggest that the ink jet process be a promising and cost-effective process fabricating high density interconnects with embedded passives in high speed electronic applications.

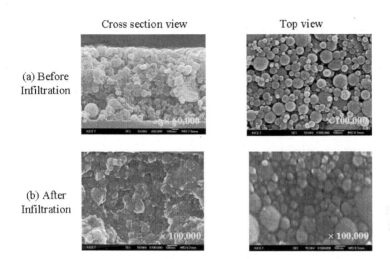

*Figure 4 The microstructure of alumina dielectrics infiltrated polymer resin via sequential ink jet process*
*: (a) before infiltration and (b) after infiltration*

CONCLUSION

The feasibility of the ink jet process for the fabrication of low loss dielectrics has been investigated. The ink chemistry and rheological properties are important in the smooth layer being printed without the clogging of the nozzle orifice. The inverse of the Ohnesorgi's number ($Z^{-1}$) can provide useful criteria for jettable ink, in which $Z^{-1}$ should be within the range of 1 to 10. The Marangoni flow, induced by appropriate mixed solvent, helps the dielectric to be flat layer suitable to high density interconnects with embedded passives. We can also fabricate low loss dielectrics consisting of highly packed alumina layer infiltrated cyanate ester resin. The Q factor of the dielectric is around 200, which can be used for high speed communication devices.

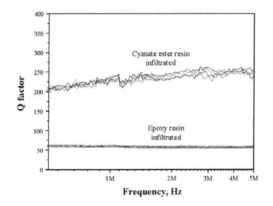

*Figure 5 The Q factors of alumina dielectrics infiltrated by polymer resins*

## ACKNOWLEDGEMENTS

This study was supported by a grant from the Fundamental R&D Program for Core Technology of Materials funded by the Ministry of Knowledge Economy, Republic of Korea.

## REFERENCES

1. T. Shimoda, K. Morrii, S. Seki and H. Kiguchi, MRS Bulletin vol. 28 (2003), p 821
2. B. Gans, P. Duineveld and U. Schubert, Adv. Mater., vol. 16 (2004), p 203
3. A. Kamyshny, M. Ben-Moshe, S. Aviezer and S. Magdassi, Macromol. Rapid Comm., vol. 26(2005), p 281
4. S. Burns, P Cain, J Mills, j. Wang and H. Sirringhaus, MRS Bulletin vol. 28 (2003), p 829
5. F. Zaugg, P. Wagner, MRS Bulletin vol. 28 (2003), p 837
6. H. Jung, S. Cho, J. Joung and Y. Oh, J. of Elec. Mater., vol. 36 (2007), 1211
7. R. Tummala, E. Rymaszewsky, A. Klopfenstein, Microelectronics Packaging Handbook Published by Springer, 1997
8. B. Derby and N. Reis, MRS Bulletin vol. 28 (2003), p 815
9. J, Moon, J. Grau, V. Knezevic, M. Cima and E. Sachs, J. Am. Ceram. Soc., 85 (2002), p 755
10. R. Deegan, O. Bakagim, T. Dupont, G. Huber, S. Nagel and T. Witten, Nature **1997**, *389*, 827
11. H. Hu and R. Larson, J. Phys. Chem., 110 (2006), 7090
12. Girard, F.; Antoni, M.; Sefiane, K.Langmuir, 24(2008); 9207
13. J. Park and J. Moon, Langmuir, vol. 22 (2006), p3506

# FORMATION OF ELECTRODEPOSITED Ni-Al$_2$O$_3$ COMPOSITE COATINGS

R. K. Saha, T. I. Khan*, L. B. Glenesk, and I. U. Haq**

Hyperion Technologies
6732 8th Street N.E.
Calgary, Alberta
Canada, T2E 7H7

ABSTRACT
This study deals with the establishment of optimum conditions for the development of wear resistant coating of Ni-Al$_2$O$_3$ composite on the steel substrates by the electrodeposition method. Each of the coating experiments was performed in an electrolytic bath, containing a dispersion of Al$_2$O$_3$ particles in nickel sulfate and boric acid solution. Composition of the coating mixture was systematically varied with respect to the contents of the dispersed particles, while the amount of the dissolved nickel sulfate, and boric acid and the applied current were kept constant during the experimental measurements. The coated substrates were characterized for their morphology, Vickers micro-hardness, and scratch resistance properties. It was observed that hardness and scratch resistance of the coated substrates increased with an increase in the Al$_2$O$_3$ content in the coating. When heat treatment at 400°C for 1h was applied to the composite coatings, a noticeable decrease in hardness was observed.

## 1. INTRODUCTION

Electrodeposition is considered one of the most important techniques for producing nano-composite coatings on the metallic substrates. The recent development and innovations in the industrial machinery has given impetus to research in this area because of its potential for growing precisely controlled wear/corrosion resistant coatings on the machine components, having complex geometry. For examples, it has been reported recently that incorporation of metallic oxides (Al$_2$O$_3$, TiO$_2$, SiO$_2$) and carbide (WC, SiC) particles[1-5] in a conventional nickel coating bath could be used as a technique for producing composite coatings on metal surfaces with a significant improvement in mechanical properties compared to a pure nickel coating. It is believed that among the essential experimental parameters for the coating process, such as pH, temperature, bath mixture composition, deposition period, etc, the hardness and size of the dispersed particles play a major role in the ultimate mechanical properties of the finished product.

In the present study, attempts have been made to develop wear resistant composite (Ni-A$_2$O$_3$) coatings on steel substrates. The Al$_2$O$_3$ ceramic powder consisted of a particle size ~500 nm in diameter and this oxide ceramic was employed as a dispersion in the Ni coating because of its high hardness value (2720 VHN).

## 2. MATERIALS AND METHODS

### 2.1. Materials

Rectangular bars (50mm x 12mm x 12mm) of carbon-manganese based steel (AISI 1018), and a nickel metal bar of the same dimensions were obtained from Metalmen Sales Inc. The Al$_2$O$_3$ powder (particle size, ~ 500 nm) was purchased from the Fischer Scientific. Analytical grade NiSO$_4$.6H$_2$O (Merck), and H$_3$BO$_4$ (Merck) were used without further purifications and the solutions and dispersions were made with distilled water.

## 2.2. Preparation of Coating Mixtures

Coating mixtures were prepared by dispersing known amounts (4 - 20 g) of the Al$_2$O$_3$ powder in 400 cm$^3$ aqueous solution, containing 1 mol dm$^{-3}$ NiSO$_4$.6H$_2$O and 0.5 mol dm$^{-3}$ H$_3$BO$_3$. These dispersions were allowed to stay for 2 h at room temperature with constant stirring before using them in the electrodeposition experiments.

## 2.3. Electrodeposition (Coating)

For this purpose, an electrolysis setup was designed in which 400 cm$^3$ of the desired coating mixture, adjusted to pH ~ 4.9, was electrolyzed at room temperature for 1 h with the DC power supply, using nickel and steel bars as anode and cathode, respectively. Both the electrodes were cleaned by sonification in acidic and basic cleaning mixtures before immersing them in the coating mixture. In all cases, the current density was maintained at 1 A dm$^{-2}$ and the content of the bath was kept agitated with magnetic stirrer during the electrolysis process. At the end of each experiment, both the electrodes were removed from the coating mixture, washed with water, dried in air, and stored in a closed container. In some cases, the coated steel bars were heated at 400 $^{\circ}$C for 1 h and then allowed to cool down to room temperature in open air.

## 2.4. Characterization

Morphology of the Al$_2$O$_3$ powder and those of the coated steel surfaces were inspected with scanning electron microscopy (SEM, JEOL JXA-8200). Coatings on the selected steel bars were also mapped with wavelength dispersive spectrometer, coupled with the same SEM.

## 2.5. Mechanical Properties

### 2.5.1 Hardness

Vickers hardness of the coated and virgin steel bars was estimated with micro-hardness tester (Leitz Mini-Load 7840) under the applied load of 50 g. In all cases, the hardness tests were conducted five times and the average value of the hardness was recorded for all the test samples.

### 2.5.2 Wear Test

The bare and selected coated steel bars were subjected to scratch tests. For this purpose, a reciprocating pin-on-disc (POD) tribometer, carrying a diamond-tipped pin as a wear generating tool, was employed as a wear tester. All the tests were conducted at room temperature for 2 minutes under the applied load of 105g for the total sliding distance of 2.4 m. The sliding velocity of the machine was kept at 0.02 m s$^{-1}$ in all the experimental runs. Optical microscope (Zeiss MC63) was used for the examination of the scratched surfaces.

## 3. RESULTS AND DISCUSSION

### 3.1. Composite Coatings

Composite coating was electrolytically deposited on the steel substrate from the aqueous dispersions, containing varying amounts (10 – 50 g dm$^{-3}$) of nanosize (~500 nm) Al$_2$O$_3$ particles, 1 mol dm$^{-3}$ nickel sulfate, and 0.5 mol dm$^{-3}$ boric acid at room temperature. It was observed that the amount of the dispersed solids in the coating mixture had significant influence on the properties of the deposited coating. For example, thickness of the coating was found to increase whereas the observed porosity therein decreased with the increase in particles loading in the coating mixture. Increase in coating thickness at high particle loading of the coating mixture was obviously due to high particles contents in the coating material. Similar trend was also reported elsewhere[6, 7] during electrolytic deposition of composite coatings on metallic substrates.

Figure 1 shows scanning electron micrographs of the coatings, obtained under the described experimental conditions. Inspection of these figures clearly reveals that the thicker coating was obtained from the mixture, composed of relatively large quantity of the dispersed Al$_2$O$_3$ particles. This observation points to the fact that in the given composition and pH of the coating mixtures, the applied current was effective enough to mobilize the positively charged[8] dispersed Al$_2$O$_3$ particles along with the nickel cations towards the cathodic substrate. Moreover, the porosity noted in the coatings, displayed in figures 1A, and C may be attributed to the evolution of hydrogen at the cathode surface, which apparently made its way through the body of the coating and thus ended with the observed porosity. It can further be seen from these micrographs that high loading of the bath mixture with Al$_2$O$_3$ particles resulted in low porosity in the obtained coating (figure 1A) as compared to the one (figure 1C) generated at low loading of the dispersed solids; possibly due to the compact structure of the coating displayed in figure 1A. Similarly, when seen from the face side (figure 1B), the coated substrate in figure 1A demonstrated grainy appearance with negligible porosity, except the void (black area) space in the centre of this micrograph which might have emerged due to the accidental damage to the coated surface. The grainy appearance of the coated surface was due to the uniformly distributed Al$_2$O$_3$ particles in the nickel matrix, the pattern of which matched well with the one seen in the micrograph (figure 2) of the dry powder of Al$_2$O$_3$. The uniformity in coating material with respect to its components (nickel, alumina, and oxygen) can also be seen from its x-ray mapping, depicted in figure 3.

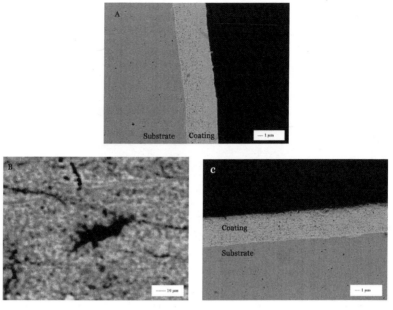

Figure 1. Scanning electron micrographs (SEM) of the coated substrates. The coating was deposited from the dispersions, 1 mol dm$^{-3}$ in nickel sulfate, 0.5 mol dm$^{-3}$ in boric acid and 50 g dm$^{-3}$ (A,B) and 10 g dm$^{-3}$ (C) in Al$_2$O$_3$ particles. Dispersion volume, 400 cm$^3$; Deposition period, 1 h; Current density, 1 A dm$^{-2}$.

Figure 2. Scanning electron micrograph (SEM) of the Al$_2$O$_3$ particles.

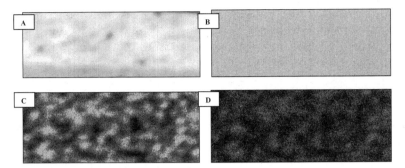

Figure 3. X-ray mapping of the Ni-Al$_2$O$_3$ coating (A) on steel substrate with respect to nickel (B), Aluminum (C), and Oxygen (D).

3.2. Micro-hardness

Figure 4 shows the micro-hardness profiles of the electrodeposited Ni-Al$_2$O$_3$ composite coatings. As can be seen from this figure, the micro-hardness of the composite coatings increased with an increase in the concentration of Al$_2$O$_3$ particles in the Ni coating. This observation was in agreement with the results reported earlier for the electrodeposition of Ni-TiO$_2$[4] and Ni-WC[9] materials onto different substrates. In fact, the large amount of Al$_2$O$_3$ particles in the coating mixture facilitated the transport of relatively large number of particles towards the cathodic substrate during the electrolytic process. Moreover, the leveling of the plot in figure 4 points to the attainment of the saturation limit for the coated layer under the applied current, mentioned in the experimental section.

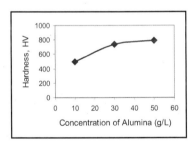

Figure 4. Micro-hardness of the coated substrate obtained in the coating mixture, containing different amounts of the dispersed Al$_2$O$_3$ particles.

Similarly, the coated samples displayed in figure 1A, was given heat treatment at 400 °C for 1 h in order to see the thermal stability of the deposited materials. Micro-hardness tests revealed that on heat treatment, the hardness of the coated sample decreased from 891 to 362 HV. The observed decrease in micro-hardness may be ascribed to the heat assisted grain growth in the deposited material, which resulted in weakening of the formed coating. In fact, our results contradict with the suggestions put forward by Krishnaveni et al[10], stating that heat treatment may enhance hardness of the coated surface due to the formation of thermally triggered pro-hardening phases in the coating matrix. This aspect of the study is further being investigated in our laboratory.

### 3.3. Wear Resistance

Wear resistance of the test samples was assessed from the width of the wear track, generated on the test surfaces in the wear experiments, described in section 2.5.2. Figure 5 shows optical micrographs of the bare steel substrate and those of the coated substrates, depicted in figures 1A and C, after subjecting them to wear tests. It can be seen from this figure that width of the wear track on the bare surface (figure 5A) was greater than those obtained on the coated surfaces (figures B, and C), which clearly indicated the weak wear resistance of the bare surface as compared to the coated surfaces. This observation indicated the fact that wear resistance of the composite coating was high as compared to the bare surface under the applied wear test conditions. Moreover, it is evident from this figure that thin coatings resulted in relatively high wear compared to the thicker coatings, which indicated that the reinforcement of the nickel matrix with a large quantity of the Al$_2$O$_3$ particles strengthened the wear resistance properties of the composite coating. Furthermore, examination of the surface morphology of the wear tracks in figure 5 indicated that the composite coatings might have undergone a change in wear mechanism. For example, the wear track in figure 5A for the uncoated surface appeared rougher than the coated surfaces shown in figures 5B and C.

The micrograph in figure 5D corresponds to a wear track produced on a coated surface subjected to heat treatment at 400°C for 1 h. Measurement taken from the wear track suggested that the heat treatment decreased wear resistance of the deposited material, possibly due to the grain-growth dependent changes in its microstructure.

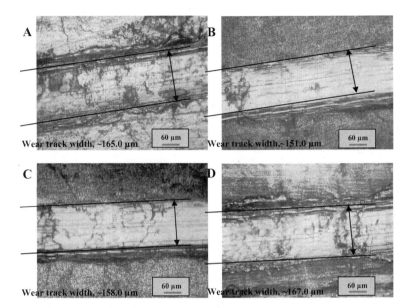

Figure 5. Optical micrographs of wear tracks, obtained with bare carbon-manganese based steel (AISI 1018) steel (A), and coated steel substrates, shown in Fig. 1A (B), and Fig. 1C (C). Micrograph in D was obtained with coated sample in Fig. 1A after heat treatment at 400 °C for 1 h. The standard deviation of the track width measurement is ±1.

## 4. CONCLUSION

1. A Ni coating containing a dispersion of Al$_2$O$_3$ was successfully deposited on the steel substrates by the electrodeposition process.

2. The applied experimental parameters, such as the amount of the dispersed particles in the bath mixture, significantly affected thickness of the deposited coating. An increase in the concentration of Al$_2$O$_3$ in the bath mixture generally led to thicker coatings.

3. The composite coatings were found to be harder and more wear resistant than the uncoated steel substrates.

4. The amount of dispersed particles of Al$_2$O$_3$ in the bath mixture proved to be an important variable in controlling the extent of porosity within the coating. Porosity was attributed to the generation of hydrogen gas at the anodic surface during the electrolytic process.

5. Heat treatment mechanically weakened the as-deposited coating, possibly due to heat assisted grain growth in its microstructure.

ACKNOWLEDGEMENTS
The authors would like to acknowledge Alberta Ingenuity Fund for the financial support for this research.

FOOTNOTES
\* University of Calgary, Department of Mechanical and Manufacturing Engineering, 2500 University Drive, N.W., Calgary, Alberta T2N 1N4, Canada.
\* \* On sabbatical leave from the National Centre of Excellence in Physical Chemistry, University of Peshawar, Peshawar 25120, NWFP, Pakistan.

REFERENCES
[1]P.A. Gay, P. Bercot, and J. Pagetti, Electrodeposition and Characterisation of Ag-ZrO2 Electroplated Coatings, Surf. Coat. Technol., 140, 147-54, 147 (2001).
[2]A. F. Zimmerman, G. Palumbo, K. T. Aust, and U. Erb, Mechanical Properties of Nickel-Silicon Carbide Nanocomposites, Mater. Sci. Eng. A, 328, 137-46 (2002).
[3]A. Grosjean, M. Rezrazi, J. Takadoum, and P. Bercot, Hardness, Friction and Wear Characteristics of Nickel-SiC Electroless Composite Deposits, Surf. Coat. Technol., 137, 92-96 (2001).
[4]J. Li, Y. Sun, X. Sun, and J. Qiao, Mechanical and Corrosion-Resistance Performance of Electrodeposited Titania-Nickel Nanocomposite Coatings, Surf. Coat. Technol., 192, 331-35 (2005).
[5]J. Steinbach, and H. Ferkel, Nanostructured Ni–Al₂O₃ Films Prepared by DC and Pulsed DC Electroplating, Scripta Mater., 44, 1813-19 (2001).
[6]I. Garcia, A. Conde, G. Langelaan, J. Fransaer, and J. Celis, Improved Corrosion Resistance Through Microstructural Modifications Induced by Codepositing SiC-Particles with Electrolytic Nickel, Corros. Sci., 45, 1173-89 (2003).
[7]Y. S. Huang, X.T. Zeng, I. Annergren, and F. M. Liu, Development of Electroless NiP–PTFE–SiC Composite Coating, Surf. Coat. Technol, 167, 207-11 (2003).
[8]D. Peak, J. Colloid Interface Sci. 303, 337-45 (2006).
[9]M. Surender, B. Basu, and R. Balasubramaniam, Wear Characterization of Electrodeposited Ni–WC Composite Coatings, Tribol. Int., 37, 743-49 (2004).
[10]K. Krishnaveni, T. S. N. Sankara Narayanan, and S. K. Seshadri, Electrodeposited Ni–B Coatings: Formation and Evaluation of Hardness and Wear Resistance, Mater. Chem. Phys., 99, 300-08 (2006).

CHARACTERIZATION OF STRUCTURES GROWN HYDROTHERMALLY ON
TITANIUM METAL FOR SOLAR APPLICATION

Judith D. Sorge and Dunbar P. Birnie, III
Materials Science and Engineering Department
Rutgers University
Piscataway, NJ USA

ABSTRACT
       The growth of titanium-based nanostructures on titanium metal creates a thick film that
may have properties that are conducive to increased efficiency in dye sensitized solar cells.
However, the growth mechanism of the structures and their composition variability is not well
understood. In this research, the composition and morphology are characterized with respect to
the growth conditions using electron microscopy and other methods. This helps us to better
understand the mechanisms of growth in a basic solution at low temperatures, less than 175°C,
for eventual incorporation into the production of dye sensitized solar cells.

INTRODUCTION
       Titanium dioxide is a material commonly used in pigments, photocatalysis and as an
ultraviolet light absorber. New structures and processing techniques have led to application in
biomimetic implants and dye sensitized solar cells (DSSC's). DSSC's are likely to soon
compete with commercially available silicon cells in cost, and, with further research, efficiency.
These next generation solar cells utilize titanium dioxide as a semiconductor into which electrons
are injected from a chemisorbed dye, which absorbs broadly in the available solar spectrum[1,2].
This research focuses on optimizing the titanium dioxide layer for morphology and electron
conduction in order to produce higher efficiency cells. The most important considerations of the
titania layer in a DSSC are high surface area and rapid electron transfer. Nanotubes of titanium
dioxide have interested researchers to fulfill these conditions, but have not yet been produced by
a simple solution based growth process[3-6].
       Kasuga, et al and other researchers have precipitated sodium titanate nanotubes from the
hydrothermal reaction of commercially available titanium dioxide nanoparticles with aqueous
sodium hydroxide[7,8]. With a variety of post-processing techniques, the sodium titanate
nanotubes can be converted to anatase titania while maintaining their structure[9-11]. However, this
method still suffers from being a solution process where dispersion of the nanotubes and their
subsequent deposition are necessary; growing the nanotubes directly on a substrate would lend
itself to ease of commercialization. This research aims to understand a similar reaction sequence
on a titanium metal substrate. Similar work was done by Tian, et al; however, as will be shown
later, a more complicated reaction sequence occurs than was previously believed[12]. Tian, et al
found that titanate nanotubes with a high surface area, normal to the surface growth pattern can
be reproducibly grown on titania coated titanium metal in a hydrothermal environment with
aqueous sodium hydroxide. However, in our work, a second plate-like morphology was found
when less titanium dioxide is provided. The conditions under which the different morphologies
grow and the mechanism of these reactions are studied in this research.

EXPERIMENTAL

Hydrothermal experiments were conducted using 0.127mm thick titanium metal (Sigma Aldrich) placed upright in a 100mL Teflon liner with 50mL of an aqueous sodium hydroxide solution. This was then placed in a stainless steel autoclave. Sodium hydroxide concentrations ranged from 1-10M and reaction times were varied from 1-24 hours. Each experiment was tested with and without the addition of 0.5g of commercially available titanium dioxide particles (Degussa, P25). After the completion of the hydrothermal reaction, the titanium metal substrate was removed and rinsed in a beaker of distilled, deionized water. Finally, the substrates were heated at 500°C for one hour; although structures were not found to change after the heat treatment, it was continued for reproducibility. Morphological characterization and energy dispersive x-ray spectroscopy (EDS) were done in the field emission scanning electron microscope (Zeiss Gemini 982). X-ray diffraction (XRD) compositional analysis was done using a Siemens D500 diffractometer.

RESULTS AND DISCUSSION

The hydrothermal reaction of titanium metal and sodium hydroxide results in two morphological structures: platelets and nanofibers. The platelets range in size from 2-20μm and the nanofibers from 10-100nm in width. Understanding the growth mechanisms that cause each of the morphologies is important to the solar cell application as only the fibers have a high surface area structure.

Figure 1. FESEM images that illustrate the different morphologies that arise in the hydrothermal reaction system. The samples above were reacted without additional titanium dioxide, in increasing concentrations of sodium hydroxide. The top row of images had a reaction time of 3 hours, and the bottom row, 18 hours.

A different reaction mechanism occurs when aqueous sodium hydroxide reacts with titanium than when titanium dioxide is the major reactant. When untreated titanium metal is hydrothermally reacted with aqueous sodium hydroxide at concentrations of 5M or greater, both platelets and wide, thin fibers form. An increase in the base concentration leads to an increase in the number of platelets in a given area, as well. After a reaction time of three hours, the platelets formed are on the order of 5μm, but these grow to an average of 20μm if the reaction continues for 18 hours. In all cases, there is a complete layer of interconnected nanofibers on the titanium substrate that surrounds the platelets. At low concentrations of sodium hydroxide, only the nanofiber layer forms and these nanofibers tend to be wider, on the order of 70-80nm. Figure 1 depicts these different structures as seen in FESEM. When titanium dioxide nanoparticles are added to the hydrothermal reaction system, only the continuous layer of interconnected nanofibers form and no platelets are seen. At higher concentrations of sodium hydroxide, the nanofibers tend towards thin, approximately 10nm, fiber strands rather than a range of widths up

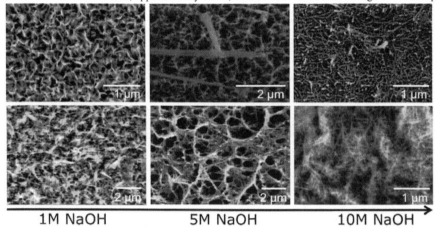

1M NaOH          5M NaOH          10M NaOH

Figure 2. The above FESEM images illustrate the nanofiber structure when titanium dioxide nanoparticles are added to the hydrothermal reaction. The top three images were reacted for 3 hours and the bottom for 18 hours, with increasing concentration of sodium hydroxide from left to right.

to 100nm, as shown in figure 2.

In order to determine the mechanisms of growth of the platelets and nanofibers, which are believed to be different phases of sodium titanate, morphology studies and characterization of the final products were completed. The nanofibers appear, from analysis of the above images, to be the product when aqueous sodium hydroxide reacts with titanium dioxide. When titanium dioxide is not in excess, platelets are formed, illustrating that they are the product of titanium metal with aqueous sodium hydroxide. Small amounts of nanofibers are still formed when no titanium dioxide is added to the reaction system because the titanium metal has a previously grown passivation layer. In these cases, the titanium dioxide is rapidly reacted, and soon after, the sodium hydroxide reacts with the pure titanium metal to form a product with platelet morphology.

EDS analysis in figure 3 demonstrates that both the nanofibers and the platelets are sodium titanate. From the differences in relative percentage of elements, it is understood that they are not of the same phase, which coincides with the conclusions from the morphological study. In order to determine the compositional phases, x-ray diffraction studies were utilized, as well as analysis of the possible chemical reactions.

In the formation of platelets, titanium metal dissolves in sodium hydroxide to form $Na_2Ti_6O_{13}$, as concluded from the XRD analysis in figure 4. The mechanistic pathway for this growth is not well understood although it is apparent that the Ti metal reacts with the sodium hydroxide to form $Ti(OH)_4$, as shown in reaction 1. Then this further reacts to form the product upon reaction with the sodium cation, with · a suggested mechanism for this reaction given in reaction 2.

When nanofibers are formed, the titanium dioxide reacts with the hydroxide anion to form a titanium oxide anion as shown in reaction 3.

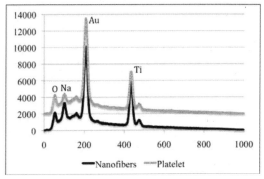

Figure 3. EDS of nanofiber and platelet morphologies grown on titanium metal in 1M aqueous sodium hydroxide at 150°C for 18 hours with no additional titanium dioxide.

Reaction 4 illustrates how the final product, $Na_2TiO_3$ is formed[13]. However, XRD analysis could not be completed on the nanofiber layer to ensure that this product is correct. Further work on this compositional analysis is ongoing.

Titanium metal and NaOH reaction:

$$Ti + 4OH^- \rightarrow Ti(OH)_4 \tag{1}$$
$$2Na^+ + 6Ti(OH)_4 \rightarrow Na_2Ti_6O_{13} + 11H_2O + H_2 \tag{2}$$

TiO₂ and NaOH reaction:

$$TiO_2 + 2OH^- \rightarrow TiO_3^{2-} + H_2O \tag{3}$$
$$TiO_3^{2-} + 2Na^+ \rightarrow Na_2TiO_3 \tag{4}$$

These conclusions have been strengthened with EDS analysis of the original products and a second set of experiments. The EDS analysis shown in figure 3 strengthens the conclusion that both the nanofibers and the platelets are a sodium titanate phase. The platelet and nanofiber compositions are seen to be quite similar. The second set of experiments utilized titanium isopropoxide instead of the addition of titanium dioxide particles. Although titanium isopropoxide hydrolyzes in the aqueous solution before the hydrothermal reaction occurs, it still forms similar intermediates and can react in a similar fashion to the titanium dioxide, though likely much more rapidly. The product is shown in figure 5 and it is apparent that it has the same nanofibrous morphology as the samples produced with crystalline titanium dioxide as the reaction source.

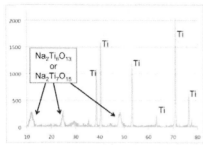

Figure 4. X-ray analysis of platelet growth on titanium metal.

Figure 5. FESEM image of titanium substrate reacted for 6 hours in 10M NaOH with the addition of titanium isopropoxide.

It should be noted that the solar cell application that we aim for depends on the subsequent conversion of the sodium titanate phase into a crystalline titania structure – hopefully while preserving the nanorod morphology. This can be done with lower sodium activity treatments that exchange $Na^+$ for $H^+$ and later heat-treat to the oxide[9]. Our current studies have not investigated that step of the reaction sequence.

CONCLUSION

It has been shown that sodium titanate can be grown on titanium metal in a hydrothermal environment using aqueous sodium hydroxide. Two distinct morphologies are formed, and the preferred structure is that of nanofibers, which creates a mesh that has high surface area for use in DSSC's. This can be reproducibly grown by adding 0.5g titania nanoparticles (for the primary Ti dissolution source in solution) to 5M sodium hydroxide and hydrothermally treating for 18hours at 150°C. It can also be produced using titanium isopropoxide as the titanium oxide precursor. It is believed that the fibrous reactant phase consists of either $Na_2TiO_3$ or $Na_2Ti_3O_7$ that is later converted to $TiO_2$. In further research, the nanofibrous mesh can maintain its high surface area even after the removal of sodium ions by using a low temperature distilled water rinse. Understanding the exact composition of both the nanofibers and the product after warm water washing is in progress through crystallographic analysis using transmission electron microscope.

ACKNOWLEDGMENTS
Support from the John Dennis/HED/Malcolm G. McLaren Fellowship is greatly appreciated.

REFERENCES

(1)    O'Regan, B.; Gratzel, M. *Nature* **1991**, *353*, 737-740.

(2)    Gratzel, M. *Journal Of Photochemistry And Photobiology C-Photochemistry Reviews* **2003**, *4*, 145-153.

(3)    Wei, M.; Konishi, Y.; Zhou, H.; Sugihara, H.; Arakawa, H. *Journal Of The Electrochemical Society* **2006**, *153*, A1232-A1236.

(4)    Flores, I. C.; de Freitas, J. N.; Longo, C.; De Paoli, M.-A.; Winnischofer, H.; Nogueira, A. F. v. *Journal of Photochemistry and Photobiology A: Chemistry* **2007**, *189*, 153-160.

(5)    Jiu, J.; Isoda, S.; Wang, F.; Adachi, M. *Journal Of Physical Chemistry B* **2006**, *110*, 2087-2092.

(6)    Yang, D.-J.; Park, H.; Cho, S.-J.; Kim, H.-G.; Choi, W.-Y. *Journal of Physics and Chemistry of Solids* **2008**, *69*, 1272-1275.

(7)    Kasuga, T.; Hiramatsu, M.; Hoson, A.; Sekino, T.; Niihara, K. *Langmuir* **1998**, *14*, 3160-3163.

(8)    Kasuga, T.; Hiramatsu, M.; Hoson, A.; Sekino, T.; Niihara, K. *Advanced Materials* **1999**, *11*, 1307-1311.

(9)    Seo, D.-S.; Lee, J.-K.; Lee, E.-G.; Kim, H. *Materials Letters* **2001**, *51*, 115-119.

(10)   Du, G. H.; Chen, G.; Peng, L. M. *Applied Physics Letters* **2001**, *79*, 3702-3704.

(11)   Tsai, C. C.; Teng, H. S. *Chemistry Of Materials* **2006**, *18*, 367-373.

(12)   Tian, Z. *Journal of the American Chemical Society* **2003**, *125*, 12384-12385.

(13)   Bavykin, D.; Parmon, V.; Lapkin, A.; Walsh, F. *Journal of Materials Chemistry* **2004**, *14*, 3370-3377.

(14)   Chen, Q.; Du, G. H.; Zhang, S.; Peng, L. M. *Acta Crystallography: Section B* **2002**, *58*, 587-593.

(15)   Ha, S. W.; Eckert, K. L.; Wintermantel, E.; Gruner, H.; Guecheva, M.; Vonmont, H. *Journal of Materials Science: Materials in Medicine* **1997**, *8*, 881-886.

(16)   Yada, M.; Inoue, Y.; Uota, M.; Torikai, T.; Watari, T.; Noda, I.; Hotokebuchi, T. *Langmuir* **2007**, *23*, 2815-2823.

ROLE OF LATTICE VIBRATIONS IN A NANOSCALE ELECTRONIC DEVICE

Karel Král
Institute of Physics, Academy of Sciences of Czech Republic, v.v.i.
Prague, Czech Republic

ABSTRACT
The electronic current through a nanotransistor is studied. The nanotransistor is assumed to consist of a quantum dot active region connected to the source and drain wires and also to a gate. The electric current is shown to be influenced by the coupling of the electrons to the longitudinal optical phonons, namely, by the up-conversion of the electrons to the higher excited states in a quantum dot, due to a nonadiabatic effect of the lattice vibrations. In the nanotransistor with asymmetric source and drain contacts the up-conversion leads to a spontaneous electric current, or to a spontaneous voltage, between the electrodes. We remind some existing experiments which might be related to the effect under consideration.

INTRODUCTION

The electronic energy relaxation in the materials of the type of GaAs is known to be strongly influenced by the interaction of electrons with the longitudinal optical (LO) phonons of the lattice vibrations [1]. This is not only true in the bulk samples but also in the low-dimensional nanostructures like quantum dots. In the latter systems the dissipation processes of the electrons are influenced not only by the strength of the electron-phonon coupling, but also by the geometrical shape of the nanostructure. This is because the charge carrier does not leave the scattering target (the LO phonons) going to infinity, but instead it reflects at the quantum dot boundary potential barriers and returns back to the target, continuing in this way the irreversible multiple scattering process. Due to the smallness of the system the efficiency of the scattering mechanism may be strongly enhanced by the multiple scattering nature of the dissipation processes.

The multiple scattering of the electrons on the LO phonons in quantum dots, included in the electron kinetic equation in the self-consistent Born approximation to the electronic self-energy, leads not only to the fast electronic energy relaxation [2,3], but also it provides the effect of an up-conversion of electronic level occupation in quantum dots [2]. The up-conversion mechanism can explain e.g. the lasing of the quantum dot lasers from the higher excited electronic states, or the photoluminescence from $SiO_2$-based nanoscale materials [4]. Recently the multiple scattering mechanism has also been shown to provide an explanation of the luminescence line shape of individual quantum dots [5], namely the form of a very narrow peak accompanied by a broader shoulder, usually at the low-energy side. Let us remark at this point that the appearence of this peculiar line shape has found a theoretical explanation so far only within the frame of the assumption of the multiple scattering of electrons on LO phonons.

Let us remark that besides the self-consistent Born approximation to the electronic self-energy in the nonequilibrium Green's function theory we utilize the instant collision approximation and the Kadanoff-Baym ansatz. The intuitive understanding of the nonadiabatic effect of the multiple scattering of the charge carriers in quantum dots, manifested by the effect of the up-conversion of electrons to the higher energy states, can be obtained as follows: in a single irreversible multiple scattering act the electron emits many virtual phonons which remain in the system of the lattice vibrations as multiphonon states. In contrast to the Fock's states of the vibrational modes of the LO phonons, the multiphonon states can remind the macroscopic coherent oscillations of the lattice, in a certain similarity to the coherent laser light, which can be approximately regarded as a classical wave of electric and magnetic field in the space. The multiphonon states then represent a certain time dependent force acting on the charge carriers. The presence of this force makes the Hamiltonian of the electronic system explicitly time-dependent, which then means that the effective electronic system is not conservative from the energy point of view. From the thermodynamics point of view the two subsystems, the electrons and the LO phonons, then do not exchange only heat. They are therefore not at an equilibrium in the usual thermodynamical sense. The appearence of the up-converted occupation of the electronic states is then not unexpected. The mechanism of the up-conversion has a similarity to the mechanism of the electron transfer in the molecular systems (R. Marcus [6]). In Marcus theory the key role is played by a certain fluctuating classical parameter, the fluctuations of which open the possibility for the electronic transfer between two molecules. The fluctuating parameter plays thus a similar role as the multiphonon vibrational state of the LO phonons does. While the theory due to Marcus gives an intuitive introduction of a key fluctuating parameter, the approach based on the excitation of the virtual multiphonon modes, resulting from the microscopic elaboration, shows that one has to go beyond the perturbation theory [7].

In an analogy with the ability of the multiple-phonon scattering mechanism to explain the fast electron energy relaxation, and the shape of the optical emission lines in quantum dots, we expect that the up-conversion mechanism plays an important role also in the open nanostructures like those of a quantum dot connected to metallic electrodes. In this paper we pay attention to the nanotransistor and show that in the case of a certain asymmetry between the source and the drain contacts such a nanodevice can display a spontaneous voltage or a spontaneous current between the contacts. We briefly turn the reader's attention to the existing experiments which are expected to be relevant in the connection with the effect under study.

NANOTRANSISTOR

In order to demonstrate the influence of the electron-LO-phonon interaction on the current-voltage characteristics of the open small system of the nanotransistor we choose a simple picture of the physical situation in this structure. The properties of the small systems under study are demonstrated on the nanotransistor consisting of a quantum dot with two nondegenerate electronic conduction band bound states, coupled to the source and drain electrodes, and exposed to the gate potential, as seen in the Fig. 1. In order to avoid the question of the positions of the chemical potentials in the quantum dot and the metallic wires, utilizing the gate electrode in the device we simply assume the electronic chemical potential of the isolated wires being put into the middle of the energy gap between the two energy levels of the isolated quantum dot.

When the contacts to the source and the drain are not symmetric then the device inclines to display a spontaneous voltage between the source and the drain, or, the device tends to display a nonzero current flowing through the device even at zero attached voltage to the source and drain contacts. This nonzero current is seen in Fig. 2. The nonzero current through the nanotransistor is enabled first of all by the up-conversion of the electronic level occupation. The nonzero electronic current at zero voltage is then made possible by the asymmetry of the contacts to the source and to the drain, due to which the electrons prefer to flow prevailingly in one direction in the device and give in this way the nonzero current even at the zero applied voltage.

The theoretical method used in this paper is based on the well-known Toy Model by S. Datta [8]. This model approach to the electronic transport in nanostructures and nanodevices gives a method of how to construct the electronic kinetic equation for the nanodevice. In fact the Toy model represents an approximation to a more general and complex theoretical method of the nonequilibrium Green's functions of Meir and Wingreen [9].

Within the approximation of the Toy Model we shall assume that the nanotransistor (Fig. 1) consists of a quantum dot which has two electronic bound states of the conduction band states, with the energies $E_A$ and $E_B$. We assume that both the energies are electronic eigenstates in the cubic shape quantum dot with the lateral size $d$, with infinitely deep three dimensional potential well, in which the electron has an effective mass $m_{eff}$. The energy $E_A$ is the electronic ground state in this potential well, while the energy $E_B$ is one of the triply degenerate lowest energy excited states. We shall confine ourselves to this two-electronic states model. We assume that in the Toy Model the two energy levels are coupled by an electronic tunneling mechanism to the two wires, $L$ and $R$. The coupling of the energy level $E_A$ to the wire $L$ will be given by a single parameter $\Gamma_{LA}$, while $\Gamma_{LB}$ will determine the tunneling efficiency of the contact $L$ to and from the energy level $E_B$. Similarly the parameters $\Gamma_{AR}$ and $\Gamma_{BR}$ have the meaning of the electronic tunneling efficiency from the level $E_A$ or $E_B$ into or from the wire $R$. We shall show that the nanotransistor displays interesting properties when the tunneling characteristics are asymmetric. For this purpose we chose $\Gamma_{LB} = \Gamma_{LA} = \Gamma_{AR} = 1$ meV, while $\Gamma_{BR} = 2$ meV. The lateral size of the dot is chosen to be $d=18$ nm. At this size the ratio between the electronic energy level separation and the longitudinal optical phonon energy is about 1.44 (the material parameters of the quantum dot material are those of GaAs).

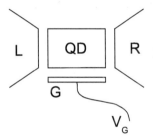

Figure 1: Schematic picture of the nanotransistor consisting of the quantum dot (QD) connected to the metallic contact $L$ and $R$. $G$ denotes the gate electrode connected to the electric potential $V_G$.

In the Toy Model the electronic transfer between the quantum dot and the electrodes is a basic part of the irreversible transfer of the charge carriers. This transfer is proportional to the broadening $2\Gamma$ of the electronic spectral density line of the quantum dot. So that the parameter $\Gamma_{LA}$ has also the meaning of the $A$-level broadening due to the virtual electronic transitions provided by the tunneling transfer Hamiltonian. Referring the reader to the original work [8] for details, the rate of the transfer of the electrons can be easily written down. For example, the electric current $I_{LA}$ from the left wire into the energy level $A$ of the quantum dot is

$$I_{LA} = \frac{-e\Gamma_{LA}}{h}(N_{LA} - N_A), \tag{1}$$

where $-e$, $e>0$, is electronic charge, $h$ is Planck constant. In the latter formula $N_A$ is the electronic occupation of the energy level $E_A$. The quantity $N_{LA}$ is the "target" occupation of the level $A$. In the Toy Model it is such an occupation which the energy level $A$ tries to achieve in order to get to a thermodynamic equilibrium with the left electrode [8].

The Toy Model describes the flow of the electrons from the left wire $L$ into the states $A$ and $B$ and from these states into the wire $R$. The Toy model does not contains a mechanism of how to make the electronic transfer between the energy levels $E_A$ and $E_B$ due to the electron-LO-phonon coupling. In the present work we simply add such a mechanism to the Toy model scheme. In the rate equation giving for example the rate of the change of the occupation of the energy level $E_A$ due to the exchange of the electrons with the wires $L$ and $R$ we just add the generation rate of electrons in that state due to the electron-LO-phonon interaction. The electron-LO-phonon relaxation mechanism between the two electronic states in the dot was studied earlier (e.g. [2,10]). In the present work the mechanism reported in these references has been generalized in a certain way. Namely, in contrast to references [2,10] the electronic energy levels in the electron-phonon relaxation rate contain the level broadening due to the exchange of the electrons between the wires and the dot, by the mechanism expressed by the parameters $\Gamma$. In this way we augment the Toy Model with the electron-phonon interaction inside the quantum dot, including the tunneling interaction between the dot and the wires. In this way we can calculate the electronic relaxation rate at the given temperature of the atomic lattice and at the given state of the electronic distribution inside the quantum dot of the nanodevice. The resulting electric current through the device depends on the electronic distribution in the quantum dot. This electronic distribution depends on the electron relaxation rate due to phonons, which in turn depends on the electronic distribution in the quantum dot. This means that the computation of the electric current flowing through the device is determined by a self-consistent procedure.

Let us also remind that the above given relaxation rate formula was computed in an approximation which corresponds to the effect of the so called phonon overheating suppressed. We refer the reader to the references [11] for the details about this shortcoming of the kinetic equations for the electronic and phonon system.

SPONTANEOUS CURRENT OR VOLTAGE

Our kinetic equations determine the electric current through the device at the given voltage attached to the electrodes $L$ and $R$. In the above outlined kinetics the energy levels of the quantum dot are input parameters, together with the attached voltage, chemical potentials the electrodes and of the

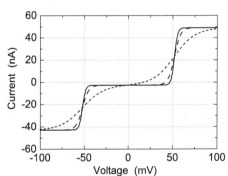

Figure 2: Dependence of the current through the device as a function of the attached voltage at the lattice temperature of 10 (full), 20 (dash) and 70 (short dash) K.

temperature of the lattice. The parameters determining the phonon assisted upconversion are hidden in the formulas [2,10] for the upconversion rate. The Fig. 2 shows that the electronic upconversion leads to a nonzero electric current of about several nA even at zero voltage applied. The electric current has a tendency to flow to the left hand side (in the Fig. 1) even at a certain range of voltage values, depending on the lattice temperature, including a range of positive values of the voltage. In other words, at these values of the voltage the electric current flows in the direction from the quantum dot to the electrode $L$, while the electrons themselves flow from the quantum dot to the electrode $R$. This is because after being upconverted to the energy level $B$ the electrons find it easier to go to the electrode $R$ due to the asymmetry in the parameters $\Gamma$. The electrons at the energy level $B$ can be exchanged between the quantum dot and the wire $R$ in a more easy way because of the increased value of the parameter $\Gamma_{BR}$.

The origin of the effect of the nonzero current can thus be explained as an implication of the asymmetry of the electric contacts and the nonadiabatic effect of the upconversion of the electrons to the level $B$. We have to keep in mind that we are obtaining this effect upon using the approximations specified above. It is seen that the electrons flowing to the right of the device get stopped at a certain value of the attached voltage, which is at the Fig. 2 equal to about 50 mV at 10 K. In practice this would mean that when the device is left alone with the contacts $L$ and $R$ disconnected from any other body and from one another, then the electrons flow to the right-hand-side electrode until the accumulated charge and the capacity of the contact $R$ create such a potential at this contact that corresponds to the critical voltage of about 50 mV shown at the Fig. 2. In the real case it should be expected that such a device will quickly capture an amount of compensating charges from its neighborhood and could become electrostatically neutralized by parasitic currents.

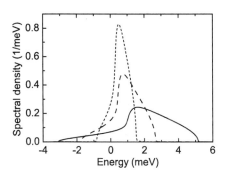

Fig. 3. The main peak of the electronic spectral density of the electronic ground state in the quantum dot without electric contacts, but with the electron-LO-phonon interaction included in the self-consistent Born approximation to the electronic self-energy. The curves refer to different lateral size of InAs quantum dot (full – 10 nm, dashed – 20 nm, short dash – 40 nm). Temperature of lattice is 10 K. Electronic distribution: 0.8 of electron is in the ground level, 0.2 electron occupies the excited state. No suppression of the overheating effect [11] is performed in the calculation.

The setup of the nanotransistor, which is, comparing to an isolated quantum dot, an open system exchanging the electrons with the neighborhood, may allow for a useful measurement of the properties of the quantum dot with varying continuously the electronic level occupation. This advantage could prove useful in confirming experimentally the dependence of the electronic spectral density on the quantum dot electronic level occupation, as it is shown in the Fig. 3. In this figure the width of the main feature of the spectral density peak is calculated depending on the change of the quantum dot lateral size. The calculated result shows that with decreasing the quantum dot lateral size, the width of the peak of the quantum dot optical transition may display an increasing trend.

COMPARISON WITH EXPERIMENTS

The measurements of the current-voltage characteristics on the devices of the nanometric size are not yet an easy experiment with individual quantum dots and individual molecules. Nevertheless, there are two experiments which may attract our attention in connection with the above presented theoretical expectations. One of these experiments concerns the measurement of the current-voltage curve on the short segments of the DNA molecule, see e.g. the experimental report [12]. We do not present the comparison of the present theory with the data given in [12], postponing these attempts to a later work. There is another experimental work reporting the measurement of the nonzero current detected on a nanotransistor with the source and the drain shunted [13]. The authors ascribe the effect to the asymmetry of the contacts, together with an upconversion of the electrons above the chemical potential of the wires caused by the thermal promotion of the electronic excited level occupation. The comparison of the thermal promotion of the excited states inside the nanotransistor with the multiple phonon scattering upconversion needs an additional attention in a future work. Clearly, more experiments are needed to compare the present theoretical results with the real device properties.

CONCLUSIONS

Using the simple model [8] of the device modelling, we present numerical results on the electronic transport in the small system of the zero-dimensional nanostructure, or a nanotransistor, in which we show the manifestation of the electron-LO-phonon interaction. Under the influence of the electronic upconversion, which is the manifestation of the nonadiabaticity caused by the multiple phonon scattering of electrons in quantum dots, and by the cooperation of the electronic upconversion with the asymmetry of the electric contacts of the nanodevice under study, the device is shown to display a tendency to spontaneously create an electric potential difference between source and drain. Alternatively, the same device, when the source and drain are shunted, is expected to generate a spontaneous electric current due to the electron-phonon coupling. A significance of the open system setup of the nanotransistor for the measurement of the properties of quantum dots with varying the electronic level occupation is emphasized.

**Acknowledgement** The support of the grant 202/07/0643 of GAČR and of the institutional project AVOZ10100520 is gratefully acknowledged.

References:

[1] D. Obreschkow, F. Michelini, S. Dalessi, E. Kapon, and M.-A. Dupertuis, Phys. Rev. B 76 (2007) 035329.

[2] K. Král, P. Zdeněk, Z. Khás, Surface Science, 566-568 (2004) 321-326.

[3] K. Král, Z. Khás: Phys. Rev. B 57 (1998) R2061-R2064.

[4] E. U. Rafailov, A. D. McRobbie, M. A. Cataluna, L. O'Faolain, and W. Sibbett, D. A. Livshits, Appl. Phys. Lett. 88 (2006) 041101; Y. D. Glinka, S.-H. Lin, L.-P. Hwang, Y.-T. Chen, N. H. Tolk, Phys. Rev. B 64 (2001) 085421.

[5] K. Král, Z. Khás, Microelectronic Engineering 51–52 (2000) 93–98; K. Král, Microelectronics Journal 39 (2008) 375–377.

[6] R. A. Marcus, Journ. Electroanalytical Chemistry 438 (1997) 251-259.

[7] K. Král, Z. Khás, phys. stat. sol. (b) 208 (1998) R5; K. Král, Z. Khás, arXiv:cond-mat/0103061.

[8] F. Zahid, M. Paulsson, S. Datta, in "Advanced Semiconductors and Organic Nano-Techniques", ed. H. Morkoc, Academic Press 2003; M. Paulsson, F. Zahid S. Datta, arXiv:cond-mat/0208183.

[9] Y. Meir, N. S. Wingreen, Phys. Rev. Lett. 68 (1992) 2512.

[10] K. Král, P. Zdeněk, Physica E 29 (2005) 341-349.

[11]  K. Král, Czechoslovak J. Phys. 56 (2006) 33; K. Král, C.Y. Lin, International Journal of Modern Physics B 22, no. 20 (2008) 3439-3460.
[12]  D. Ullien, H. Cohen, D. Porath, Nanotechnology 18 (2007) 424015.
[13]  Horsell, A. K. Savchenko, Y. M. Galperin, V. I. Kozub, V. M. Vinokur, D. A. Ritchie, Europhys. Lett., 71 (2005) 658.

# MODIFICATION OF QUARTZ FABRIC WITH MULTI-WALLED CARBON NANOTUBES FOR MULTIFUNCTIONAL POLYMER COMPOSITES

A. N. Rider[1], E. S-Y. Yeo[1], N. Brack[2], B. W. Halstead[2] and P. J. Pigram[2]
[1]Defence Science and Technology Organisation, Victoria, Australia, 3207.
[2]Centre for Materials and Surface Science, Department of Physics, La Trobe University, Victoria, Australia, 3086.

ABSTRACT
Multi-walled carbon nanotubes (MWCNTs) have been grown onto plain-weave quartz fabric using chemical vapour deposition (CVD). The MWCNT coated quartz was then infused with either epoxy or bismaleimide (BMI) resins to produce composite laminates. It was anticipated that the MWCNT modified laminates may exhibit improved mechanical and electrical properties compared with the base laminate. The MWCNT coated quartz fabric was heat treated and plasma modified prior to resin infusion. The laminates were characterised and their mechanical, thermal and electrical properties measured. MWCNTs grew in a quasi-perpendicular direction on individual fibre surfaces, which comprised the woven strands of the fabric, subsequently the coated quartz mats were significantly thicker than the uncoated fabric. Some improvement in the mode I fracture toughness for epoxy based composites modified with MWCNTs was observed. BMI laminates modified with MWCNTs showed good electrical and thermal conductivity but decreases in modulus and flexural strength. Poor adhesion between MWCNTs, the quartz fibres and resin may have contributed to the reduced mechanical performance.

## INTRODUCTION

Recent research into hybrid composite materials using carbon nanotubes (CNTs) has examined methods to modify the fibre reinforcement with CNTs prior to infusion of the matrix phase [1-5]. The greatest improvement in mechanical properties was reported for CNT-modified plain weave silicon carbide fabric infused with epoxy resin [1]. Increases in flexural strength and mode I fracture toughness of 250% and 350%, respectively, were reported. The growth of CNT forests perpendicular to the fibre direction also improved the through-thickness electrical and thermal conductivity. A study of alumina fabric modified with CNTs prior to infusion with epoxy resins demonstrated a 30% improvement of interlaminar shear strength and large increases in electrical conductivity compared to the base laminate [2]. The production of hybrid composites using carbon fibres initially did not show the same levels of improvement in mechanical properties as composites using ceramic fibres. Early work examining the pull-out strength of CNT coated carbon fibres in epoxy showed a 15% increase in interfacial strength, whereas heat treatment of the fibre without CNT growth decreased the strength by 30% [3]. In a recent study investigating chemical vapour deposition (CVD) growth of CNTs onto plain-weave carbon fibres prior to phenolic resin infusion, 75% and 54% increases in flexural strength and modulus, respectively [4], for the composites were reported. Further studies using electrophoresis to deposit CNTs onto carbon fibres in place of CVD processes showed a 30% improvement in interlaminar strength and through-thickness electrical conductivity [5].

This paper investigates modification of plain-weave quartz fabric by CVD growth of MWCNTs prior to infusion with either epoxy or BMI resin. The MWCNT coated fabric was heat treated and plasma modified prior to infusion. Laminates were characterised to establish distribution of the MWCNTs in the matrix phase and to establish fibre, matrix and MWCNT volume fractions. The mechanical, electrical and thermal properties of the laminates were compared with laminates prepared from unmodified fabrics. Failure analysis was undertaken to establish fracture processes in the MWCNT modified laminates.

EXPERIMENTAL

Plain-weave Astroquartz® II fused silica fabric (style 525) was purchased from JPS Composites with a fabric density of 67.8 g/m². Each yarn consisted of 120 filaments of 9μm diameter. Slow crystallisation or devitrification was reported to occur above 1050°C [6].

The quartz tube reactor used in the production of MWCNTs (Figure 1) had an 11.5 cm diameter and a uniform heating zone of approximately 40 cm in length. Quartz fabric (34 cm x 24 cm) was clamped to a metal frame prior to insertion in the reactor. MWCNTs were deposited onto the quartz fabric by catalytic chemical vapour deposition of a xylene-ferrocene feedstock in an argon/hydrogen atmosphere at 700-800°C [7]. The MWCNTs grown in the reactor under these conditions typically had a mean diameter and length of 55 nm and 55 μm, respectively. The quantity of MWCNTs grown onto the fabric was estimated by weighing small coupons placed throughout the reaction zone.

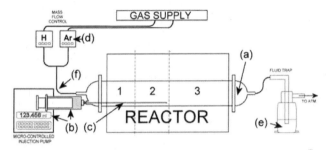

Figure 1. A diagrammatic representation of the scaled-up CVD reactor indicating the individual components (a) Glass reactor tube (b) Injection pump and syringe (c) Hypodermic needle (d) Mass flow controllers (e) Fluid trap and (f) Gated O₂ trap.

Coated fabric was heat treated at 420°C for 15 minutes and plasma treated using an Atomflo® 500 atmospheric plasma unit from Surfx Technologies at 140W to 200W at a rate between 2.5 mm/s to 5.0 mm/s using 30 L/min of helium and 0.5-1.5 L/min of oxygen from a standoff distance of 2.5 mm. The coated mats were characterised using XPS and SEM. XPS experiments were performed using a Kratos Axis Ultra spectrometer with a Al $K_{\alpha 1}$ (1486.6 eV) X-ray source operated at 150 W. SEM imaging used a LEO 1530VP SEM operating at 2 kV.

Preparation of the composite laminates used either LY5052/HY5052 epoxy resin or 5250-4 RTM bismaleimide (BMI) resin. Quartz/epoxy laminates were prepared using wet lay-up procedures and cured under 275 kPa autoclave pressure at 30°C followed by a postcure for 8 hours at 80°C. Quartz/BMI laminates used powder resin film infusion to produce prepreg material. The BMI RTM resin powder was weighed to provide the required fibre volume fraction and heated at 85°C and -100 kPa vacuum pressure to infuse the CNT coated quartz mat. The prepreg material was then cut to size, vacuum bagged and cured at 177°C and 586 kPa for 6 hours, followed by free standing post cure at 226°C for 6 hours. Details of the prepared laminates and testing are provided in Table I.

Table I. Details of the modified quartz laminates prepared for testing

| Description | Details | Testing |
|---|---|---|
| Quartz/Epoxy | 15 plies as-received quartz laminate. | 3 point bend |
| | 6 plies quartz laminate bonded to metal adherend with toughened epoxy | mode I test |
| Quartz-SR/Epoxy | 15 plies laminate with sizing removed by heat treatment at 500°C/2hours. | 3 point bend |
| Quartz/CNT/Epoxy | 6 plies CNT coated fabric interleaved with 9 plies quartz fabric | 3 point bend |
| | 2 plies CNT coated fabric laminate bonded to metal adherend with toughened epoxy | mode I test |
| Quartz/CNT-H/Epoxy | 6 plies CNT coated fabric heat treated and interleaved with 9 plies quartz fabric | 3 point bend |
| Quartz/BMI | 24 plies laminate. Quartz mat heat treated 2 hours at 800°C in Ar/$H_2$ prior to layup. | 4 point bend/ short beam shear/ electrical and thermal conductivity |
| Quartz/BMI-RR | 16 plies heat treated quartz infused with excess resin | |
| Quartz/CNT/BMI | 12 plies Heat treated CNT coated quartz laminate | |
| Quartz/CNT-Pl/BMI | 12 plies Heat and Plasma treated CNT coated quartz laminate | |
| Quartz/CNT/BMI-RR | 12 plies Heat treated CNT coated quartz infused with excess resin | |

Laminates were cut to size and tested using the short beam shear test (ASTM D2344), the mode I test (ASTM D 5528-01), the 4 point flexure test (ASTM D 6272 – 02) and the 3 point flexure test (ASTM D 790 – 02). In the case of the mode I testing, 2 to 6 plies laminates were secondarily bonded to metal backed adherends using a rubber toughened epoxy adhesive. The mode I fracture toughness was calculated using the modified beam theory. Fracture analysis of the failed laminates was undertaken using LEO 1530VP SEM, with quartz samples requiring gold coating.

Fibre volume fractions of the laminates were determined by measuring the specific gravity of the composites in water (ASTM D 792-00) followed by burning the matrix material off at 600°C to determine the fibre weight (ASTM D 2584 – 02). Laminates were also potted and polished in cross-section and analysed with a Hitachi 5200 In Field - Field Emission SEM. All samples were coated with a 3.5 nm layer of ion beam-sputtered iridium. Energy dispersive spectroscopy (EDS) point scans across a CNT coated fibre-matrix interface were also acquired to establish levels of iron at the fibre interface.

Electrical conductivity of the cured laminates was undertaken using the standard 4 point probe method with samples of 40 mm by 12.7 mm dimensions. Thermal resistance (R) of 50 mm by 50 mm specimens was measured with a KES-F Thermolabo II using equation 1.

$$R = \frac{(T_{hot} - T_{cold}).A}{W} \qquad (1)$$

Where A was the surface area of the hot plate ($25 \times 10^{-4}$ m$^2$), $T_{cold}$ was the temperature of the cold plate under the sample, $T_{hot}$ was the temperature of the hot plate on top of the sample and W was the heat flow. Thermal conductivity was calculated using R and sample thickness.

RESULTS
Characterisation of MWCNT Coated Fabrics
Figure 2 shows low and high magnification images of quartz and CNT coated fabric, indicating significant coverage of MWCNTs. Typical weight increases in quartz fabric are shown in Table II as well as the effect of heat treatment of the coated fabric. After 15 minutes of heat treatment the weight increase in fabric due to MWCNTs was approximately 25%.

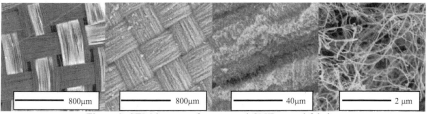

Figure 2. SEM images of quartz and CNT coated fabric.

Table II. Weight of CNTs grown on quartz fabric before and after heat treatment.

| Relative CNT weight | Heat treatment time at 420°C (min) | | | |
|---|---|---|---|---|
| (% of fabric) | 0 | 5 | 10 | 15 |
| Average | 34.1 | 34.0 | 30.6 | 27.2 |
| Std. Dev. | 3.2 | 3.2 | 2.9 | 2.6 |

Table III and Figure 3 show the peak fitting results from XPS analysis of the CNT coated quartz fabric before and after heat and plasma treatment. The peak fitting used HOPG graphite as a reference peak and adjusted the C 1s spectra of the CNT coated samples to determine relative changes caused by heat and plasma treatment. The results suggest that heat treatment did not substantially alter the chemical composition of the CNT surface, however, the plasma treatment increased the C-O type species by around 5% and decreased the graphitic component.

Table III. Atomic concentration of carbon species and oxygen for CNT coated quartz fabrics

| | Heat Treatment (min) | 0 | 20 | 20 |
|---|---|---|---|---|
| | Plasma Treatment (min) | 0 | 0 | 2 |
| Concentration (Atomic%) | graphite | 80.4 | 80.7 | 72.7 |
| | C-H | 14.9 | 12.6 | 9.8 |
| | C-O | 2.6 | 2.5 | 7.5 |
| | C=O | 0.6 | 1.9 | 1.9 |
| | O-C=O | --- | --- | 0.7 |
| | O | 0.6 | 1.3 | 7.2 |

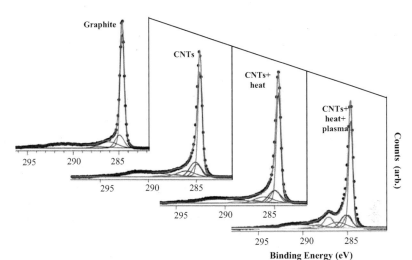

**Binding Energy (eV)**

Figure 3. XPS peak fitting of CNT coated quartz fabric, before and after heat and plasma treatment compared to graphitic reference.

Characterisation of Composite Laminates

The average CNT coated quartz ply thickness was around 2.5 times greater than the untreated quartz ply (Table IV) as determined from optical cross-sections of Quartz/Epoxy and Quartz/CNT/Epoxy laminates (Figure 4). The results showed that the 6 plies of CNT coated quartz fabric substantially increased the thickness of the 15 plies laminates, although heat treatment appeared to improve laminate consolidation through improved wetting.. The optical images showed that the space between individual fibres in the strand was as great as 20 µm, whereas the quartz laminate fibres were tightly bundled.

Table IV. Thickness of the interleaved CNT coated plies in the Quartz/Epoxy laminate.

| Interleave | Fabric thickness (µm) | |
|---|---|---|
| layer no. | CNT | CNT +Heat |
| 1 | 178 | 250 |
| 2 | 104 | 126 |
| 3 | 377 | 116 |
| 4 | 172 | 128 |
| 5 | 99 | 254 |
| 6 | 245 | 253 |
| average | 196 | 188 |

SEM images of Quartz/BMI and Quartz/CNT/BMI laminates are shown in Figure 5. The CNT coated quartz laminates indicated a concentration of CNTs in close proximity to the fibre surface, which diminished as the distance away from the fibre increased. The Quartz/CNT/BMI image showed some fine pinholes near the fibre-matrix interface that were not present for the

Quartz/CNT-Pl/BMI sample, suggesting that plasma treatment might have improved wetting of the CNT surface by the BMI resin during infusion and cure. The images also showed cracks between the fibre-matrix interface for the untreated quartz laminates that were not evident in the CNT treated quartz laminates, possibly indicating resin toughening at the fibre-matrix interface by the CNTs. An EDS line scan through a CNT coated fibre was also acquired (Figure 6). The iron concentration appeared to increase before the edge of the fibre on each side, suggesting that some surface diffusion of the iron catalyst might have occurred during CNT deposition.

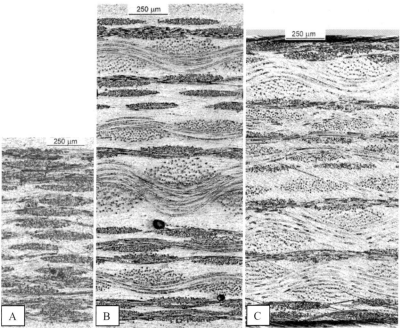

Figure 4. Optical cross-sections of 15 plies laminates: (A) Quartz/Epoxy, (B) Quartz/CNT/Epoxy and (C) Quartz/CNT-H/Epoxy

Thermal and Electrical Testing

Table V shows the electrical and thermal conductivity measured for the quartz and CNT modified Quartz/BMI laminates. The results indicate a large increase in electrical conductivity for the CNT modified laminates compared with the insulating Quartz/BMI laminates. The thermal conductivity of the CNT modified laminates also increased by around 150% compared to the Quartz/BMI laminate.

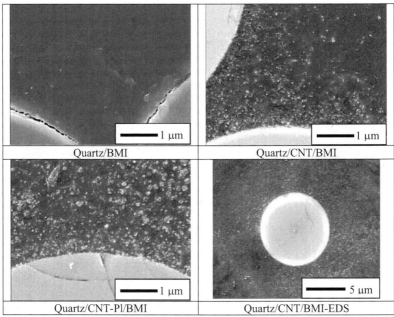

Figure 5. SEM images of the Quartz/BMI, Quartz/CNT/BMI and Quartz/CNT-Pl/BMI laminates polished in cross-section and an image showing positions where EDS point scans were acquired.

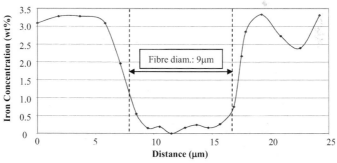

Figure 6. EDS point analysis through a quartz fibre coated with CNTs and infused with BMI.

Mechanical Testing

Table VI shows the 3 point bending strength and mode I fracture toughness values measured for Quartz/Epoxy laminates. Within the spread of results all strength values are similar apart from the Quartz/Epoxy laminate where the quartz fabric had the sizing burned off prior to resin infusion. The data suggests that the CNT modified fabric layers might have had similar properties to the fabric layers with sizing. The mode I fracture toughness values almost doubled for the CNT modified fabric, suggesting that significant improvement in the matrix fracture toughness resulted from the addition of CNTs.

Table VII shows the 4 point bending modulus, strength and short beam shear strength for the Quartz/BMI laminates. Table VII also includes the volume fractions of the laminate constituents. The results show a strong correlation in fibre volume fraction with the flexural strength, modulus and short beam shear strength. The relative change in short beam shear strength at 177°C compared to room temperature was similar for quartz and CNT modified laminates, suggesting that no obvious changes in the thermo-mechanical properties of the CNT modified matrix occurred.

Table V. Electrical conductivity and thermal resistance for Quartz/BMI laminates

| Laminate | Thickness (mm) | Conductivity (S/cm) | Thermal Resistance $(K.m^2W^{-1})$ | Thermal Conductance $(W\ m^{-1}\ K^{-1})$ |
|---|---|---|---|---|
| Quartz/BMI | 1.8 | --- | 0.0068 | 0.26 |
| Quartz/CNT/BMI | 1.5 | 2.0 | 0.0039 | 0.38 |
| Quartz/CNT-Pl/BMI | 1.9 | 1.8 | 0.0048 | 0.40 |

Table VI. Modulus and strength for Quartz/Epoxy laminates tested in 3 point bending

| Sample | Ply Thickness (μm) | | $\sigma_b$ (MPa) | $G_I$ (J/m$^2$) |
|---|---|---|---|---|
| | Laminate | CNT plies | | |
| Quartz/Epoxy | 75 | --- | 347 ± 17 | 477±43 |
| Quartz-SR/Epoxy | 75 | --- | 212 ± 12 | --- |
| Quartz/CNT/Epoxy | 137 | 196 | 314 ± 24 | 910±133 |
| Quartz/CNT-H /Epoxy | 122 | 188 | 314 ± 18 | --- |

Fracture Analysis

SEM images of the fractured Quartz/Epoxy and Quartz/BMI laminates are shown in Figure VIII. The Quartz/CNT/Epoxy surface showed roughened fracture morphology where the crack moved between the CNT and matrix interface. The Quartz/BMI surface indicated a smooth featureless appearance, indicative of brittle fracture. The CNT modified surface showed brittle fracture in the resin rich regions, whereas in the areas where CNTs were exposed, they were pulled out of the matrix. The plasma treated CNT modified surface displayed a more textured fracture surface and evidence of CNTs and resin still attached to the glass fibres. All the other fracture surfaces showed relatively clean glass fibre surfaces, although some residual matrix and CNTs were present on the Quartz/CNT/BMI glass fibres.

DISCUSSION

Catalytic chemical vapour deposition provided an efficient method of coating quartz fabric with MWCNTs. SEM (Figure 2) and optical images (Figure 4) suggest that a dense forest of CNTs grew on individual fibres of the quartz strands and, on average, individual plies in the laminates were as much as twice the thickness of untreated plies. XPS analysis of the CNT coated fabric (Table IV) suggested that there might be improved resin wetting afforded by the additional oxygen functionalities on the CNT surfaces provided by the plasma treatment. SEM images of the interfacial region between the quartz fibre and resin matrix indicated that the plasma treated CNTs were free of micro-voids present for the heat treated CNTs (Figure 5). The SEM images of the interfacial regions of CNT modified laminates (Figure 5) indicated a heterogeneous distribution of CNTs. At the fibre/matrix interface, there was a high density of CNTs which decreased further into the matrix. There was some indication that iron enrichment of the outer fibre surface might have occurred where the catalyst from the CVD process would be deposited at high temperature, potentially modifying the quartz chemistry in this region.

Table VII. Modulus and strength for Quartz/BMI laminates tested in 4 point bending (4pt) and short beam shear (sbs) for volumes of fibre ($V_f$), matrix ($V_m$), CNT ($V_{CNT}$) and iron ($V_{Fe}$)

| Sample | Ply Thick. ($\mu$m) | $V_f$ (%) | $V_m$ (%) | $V_{CNT}$ (%) | $V_{Fe}$ (%) | Test | Test Temp. | $E_b$ (GPa) | $\sigma_b$ (MPa) |
|---|---|---|---|---|---|---|---|---|---|
| Quartz/ BMI | 73 | 42.3 | 57.5 | --- | --- | 4pt | RT | 20 ± 2 | 364 ± 20 |
| | | | | | | sbs | RT | --- | 47 ± 3 |
| | | | | | | | 177°C | --- | 41 ± 4 |
| Quartz/ BMI-RR | 93 | 32.7 | 67.3 | --- | --- | 4pt | RT | 18 ± 2 | 273 ± 19 |
| | | | | | | sbs | RT | --- | 23 ± 2 |
| Quartz/ CNT/BMI | 125 | 24.6 | 66.6 | 10.0 | 0.7 | 4pt | RT | 12 ± 2 | 212 ± 7 |
| | | | | | | sbs | RT | --- | 27 ± 3 |
| | | | | | | | 177°C | --- | 20 ± 5 |
| Quartz/ CNT-Pl /BMI | 158 | 18.6 | 77.6 | 7.4 | 0.7 | 4pt | RT | 11 ± 2 | 154 ± 15 |
| | | | | | | sbs | RT | --- | 21 ± 5 |
| | | | | | | | 177°C | --- | 16 ± 2 |
| Quartz/CNT /BMI-RR | 392 | 7.8 | 90.6 | 2.1 | 0.6 | 4pt | RT | 7 ± 1 | 92 ± 16 |
| | | | | | | sbs | RT | --- | 12 ± 1 |

The good coverage of CNTs and apparently good laminate consolidation were reflected in significant improvements in both electrical and thermal conductivity (Table V), which would suggest that a continuous network of CNTs had formed throughout the laminate. The electrical conductivity was 5 times better than that reported for CNT-silicon carbide hybrid laminates but slightly lower than the thermal conductivity [1]. As thermal conductivity of the hybrid composite depends partly on the quality of the CNTs [8], this may indicate that CNTs produced in this work had more structural defects than in reference [1]. The thermal conductivity of the CNT modified Quartz/BMI laminates was approximately half that reported for single walled CNT powder added to epoxy resin at substantially lower volume fraction, but the electrical conductivity was 20 times higher [14]. Compared with MWCNT powder added to epoxy resin at volume fractions above percolation threshold, the CNT modified quartz/BMI laminates had thermal conductivity ~ 30% higher and electrical conductivity ~ 300 times higher [9]. The volume fraction of MWCNTs contained in the BMI matrix in the current work was around 15% and 10% volume fraction for the heat treated and plasma modified CNT based laminates, respectively, which was substantially greater than volume fractions of CNTs powders required to achieve percolation threshold in epoxy resin [8,9]. The similar electrical and thermal performance of the plasma treated CNT coated fabrics to the heat treated CNT coated fabrics suggests that partial oxidation of the CNT outer graphitic structure did not significantly impact on the electrical or thermal properties of the CNT modified Quartz/BMI laminates.

| | |
|---|---|
| 2μm | 20μm |
| Quartz/CNT/Epoxy | Quartz/BMI |
| 10μm | 10μm |
| Quartz/CNT/BMI | Quartz/CNT-Pl/BMI |

Figure 8. SEM failure images from Quartz/Epoxy and Quartz/BMI flexure specimens

Despite the good electrical and thermal properties measured for the CNT modified Quartz/BMI laminates, the mechanical strength showed deterioration from the baseline laminate with the heat treated quartz. Flexural modulus and strength for Astroquartz® / polyimide laminates of 22 GPa and 700 MPa [6], respectively, compared to equivalent values of 20 GPa and 364 MPa for the Quartz/BMI laminate, indicating that the heat treatment at 800°C had a significant impact on strength. Further understanding of the differences in modulus and strength values for the Quartz/BMI and CNT modified laminates is provided by plots in Figure 9. The modulus and strength of the laminates is plotted as a function of the fibre volume fraction and the data is fitted according to the rule of mixtures for composite materials [10]. Plain weave fabric was modelled as unidirectional fibres assuming the fibre volume fraction was only 50% of the measured value (Table VII). CNT modified laminates had the matrix modelled as a random short fibre composite using the CNT volume fractions from Table VII. The CNT aspect ratio was assumed to be 20 to allow for short tubes between fibre bundles and the modulus was assumed to be 100 GPa. Laminate failure was assumed to occur at the fibre failure strain.

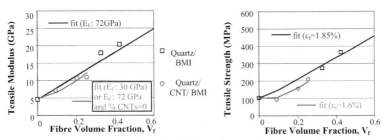

Figure 9. Plots of Quartz/BMI laminate modulus and strength indicating fits to the data using rule of mixtures theory [10].

Satisfactory fits for modulus of CNT modified laminates were only achieved by assuming the quartz fibre modulus was 30 GPa, which was less than half the value quoted by the manufacturer and that used to fit the Quartz/BMI laminates [6]. Alternatively, the CNTs could be considered as voids and 72 GPa could then be used to model the fibre modulus to achieve a similar fit to that shown in Figure 9. The modelling of the modulus suggests that either the CVD process has affected the fibre modulus or that bonding between the resin and CNTs is absent. The absence of CNT to fibre bonding may also be important. CNTs did not appear to be well adhered to the fibre surface as shown in the fracture surfaces (Figure 8), which contrasted with silicon carbide-CNT laminates, where the CNT coated fibres were annealed at high temperatures to improve the adhesion strength [11].

The tensile strength was estimated from the laminate modulus and the best fit strain to failure values. The fitted and measured strain values were similar for the quartz and CNT modified quartz laminates. The average failure strain for the CNT modified laminates was 85% of the Quartz/BMI laminates. Iron absorption at the quartz surface (Figure 6) may have increased the number of surface flaws, leading to a reduction in fibre stiffness and strength. The failure strain values of 1.9% and 1.6% for the Quartz/BMI and Quartz/CNT/BMI laminates were significantly below 3.3% strain value modelled for Astroquartz® / polyimide laminates [6]. There is indirect evidence to suggest that the CVD process led to weakening of the Astroquartz® fibres. Further work is required to examine the properties of laminates when the CNTs have been burnt off the quartz fabric before resin infusion, as this should provide a more direct measure of the CVD impact on fibre properties.

Mode I testing of the Quartz/CNT/Epoxy laminates indicated that the CNTs were capable of improving fracture toughness. This result may suggest that for testing where matrix properties dominate and load is applied in the direction closer to the general CNT orientation then some improvement in composite properties would be possible.

CONCLUSIONS

The following conclusions regarding CNT modified of quartz/epoxy and quartz/BMI laminates can be made:

1) CNTs grew in a quasi perpendicular direction to the individual fibre surface and increased the fabric thickness by two to three times.

2) Heat and plasma treatment of the CNT fabric improved resin wetting.

3) CNT distribution was concentrated close to the fibre surface and decreased between plies, even for well consolidated laminates.

4) Coating the quartz fabric with CNTs prior to infusion enabled production of relatively high volume fraction CNT laminates with good electrical and thermal conductivity

5) Work to characterise the mechanical properties of quartz fabric after removal of CNTs is required to help determine the reasons for a decrease in the mechanical properties of the CNT modified laminates.

REFERENCES

[1] V. P. Veedu, A. Cao, X. Li, K. Ma, C. Soldano, S. Kar, P. M. Ajayan and M. N. Ghasemi-Nejhad, Multifunctional Composites using Reinforced Laminae with Carbon-Nanotube Forests, *Nat Mater,* **5**, 457-462 (2006).

[2] E. J. Garcia, B. L. Wardle, A. J. Hart, N. Yamamoto, Fabrication and Multifunctional Properties of a Hybrid Laminate with Aligned Carbon Nanotubes Grown In Situ, *Compos Sci Technol,* **68**, 2034–2041 (2008).

[3] E. T. Thostenson, W. Z. Li, D. Z. Wang, Z. F. Ren, T. W. Chou, Carbon Nanotube/Carbon fiber Hybrid Multiscale Composites, *J Appl Phys*; **91**, 6034–6037 (2002).

[4] R.B. Mathur, S. Chatterjee, B.P. Singh, Growth of Carbon Nanotubes on Carbon Fibre Substrates to Produce Hybrid/Phenolic Composites with Improved Mechanical Properties, *Compos Sci Technol* , **68**, 1608–1615 (2008).

[5] E. Bekyarova, E. T. Thostenson, A. Yu, H. Kim, J. Gao, J. Tang, H. T. Hahn, T.-W. Chou, M. E. Itkis and R. C. Haddon, Multiscale Carbon Nanotube-Carbon Fiber Reinforcement for Advanced Epoxy Composites, *Langmuir*, **23**, 3970-3974 (2007).

[6] Astroquartz® Product Manual, JPS Composite Materials, PO Box 242 Slater Road Slater, SC 29683, jpscm.com.

[7] B. W. Halstead, N. Brack, A. N. Rider, E. Yeo, P. J. Pigram, in *2006 International Conference on Nanoscience and Nanotechnology*, **Vol. 1** (C. Jagadish and G. Q. M. Lu, eds.), IEEE Nanoscience and Nanotechnology, p. 146 (2006).

[8] A. Yu, M. E. Itkis, E. Bekyarova and R. C. Haddon, Effect of Single-Walled Carbon Nanotube Purity on the Thermal Conductivity of Carbon Nanotube-Based Composites, *Appl Phys Letts*, **89**, 133102 (2006).

[9] A. Moisala, Q. Li, I.A. Kinloch, A.H. Windle, Thermal and Electrical Conductivity of Single- and Multi-Walled Carbon Nanotube-Epoxy Composites, *Compos Sci Technol,* **66**, 1285–1288 (2006).

[10] P. K. Mallick, Fiber-Reinforced Composites-Materials Manufacturing and Design, 2$^{nd}$ Ed., Chapter 3, pp. 91-191, Marcel Dekker, New York, 1993.

[11] A. Cao, V. P. Veedu, X. Li, Z. Yao, M. N. Ghasemi-Nejhad and P. M. Ajayan, Multifunctional Brushes made from Carbon Nanotubes, *Nat Mater,* **4**, 540-545 (2005).

# FABRICATION OF SILICON-BASED CERAMIC SYNTHESIZED FROM MESOPOROUS CARBON-SILICA NANOCOMPOSITES

Kun Wang[1], Yi-Bing Cheng[1]
[1]Department of Materials Engineering, Monash University, Clayton Campus, VIC 3800 Australia
Huanting Wang[2]
[2]Department of Chemical Engineering, Monash University, Clayton Campus, VIC 3800 Australia

## ABSTRACT

Mesoporous carbon-silica ($C\text{-}SiO_2$) nanocomposites with different $C/SiO_2$ molar ratios were used as precursor for fabricating silicon-based ceramics. Different silicon carbide nanostructures were synthesized by carbothermal reduction of mesoporous $C\text{-}SiO_2$ nanocomposites via different heat treatments under argon. X-ray diffraction, scanning electron microscopy, transmission electron microscopy, and nitrogen adsorption-desorption analysis were used to characterize $C\text{-}SiO_2$ nanocomposites and SiC products. The major morphologies formed from the mesoporous $C\text{-}SiO_2$ nanocomposites were nanoparticles and nanofibers. With higher quantity of P123, which is the surfactant for the mesopores, the BET surface area and pore volume increased, inducing a decrease of nanofibers. The mesoporous precursors were also heated at 1200 °C for 15 hours to make a nearly dense structure and then heated to the final temperature. The products were almost nanoparticles which had a larger size than those heated directly to 1450 °C. Therefore carbothermal reduction of mesoporous $C\text{-}SiO_2$ precursors appears to be an effective means of accelerating the reaction and controlling SiC nanostructures.

## INTRODUCTION

The first ceramic materials were probably earthenware. The process of preparing these materials was quite simple. Soil which contains "organic binder" (humus) was mixed with solvent and shaped into a green body. The green body was dried and fired to obtain ceramic pots. The next materials were glasses which were shaped into objects in the molten state – an idea similar to metallurgy[1]. This can be expected because silica is among the most common naturally occurring materials. High-purity silica-based materials can be prepared by purification of naturally occurring materials, oxidation of high-purity elements, or the hydrolysis of inorganic salts and alkoxides.

The methods of preparation of most modern non-oxides ceramics can be tracked back to the mid 1800's. The majority of these materials were prepared by direct reaction of elements or by carbothermal reduction of metal oxides. A third method was from the halides of elements. The industrial processes for the large scale production of materials are also based on these methods. For example, silicon carbide is commercially prepared from sand and coke by a carbothermal reduction process[2]. Even today, this is the most economical method of the preparation of silicon carbide. Although these methods have now been known for 150 years, research still continues.

Silica sources include a wide variety of materials such as rice hulls, wheat husks, colloidal silica, silica gel, sea sand and hydrothermal water. Carbon sources include carbon black, charcoal and gaseous hydrocarbons. All of the reported Si/C ratios used excess carbon in the reaction with the ratios ranging from 0.6 to 3 (by weight)[3]. The reactions have been reported to occur in several atmospheres including nitrogen, argon, carbon monoxide, hydrogen and combinations of these[4].

Carbothermal reduction reactions using oxide-based precursors form the basis for nearly all of the commercial and high volume production of SiC whiskers in the world[5]. As such, the range of materials used for precursors and the reaction conditions used are virtually unlimited. In the process, both SiO and CO gases are produced. While the SiO is an intermediate, the CO is a by-product that must be removed from the system. At high concentrations of CO, the production of SiO is inhibited. Another consideration in the processes is that SiO gas is "heavy" and tends to segregate in systems using low molecular weight gases (like hydrogen and nitrogen) or in closed containers. High SiO concentrations have been observed to result in irregular growth morphology of the SiC whiskers[3]. Thus, reactor design to maintain the proper conditions during whisker growth is a major component in the production process.

In addition to whisker growth, considerable quantities of SiC particulates are also produced during carbothermal reduction. Excess carbon is typically added to the silica to ensure complete reaction. Any unreacted portion remains in the product material that comes directly from the reactor. In fact, reported contents from a large scale whisker growth reaction list the products as 15% SiC whiskers, 60% SiC particulates and 25% residual carbon[6].

The preparation of inorganic materials by thermolysis of preceramic compounds has become of substantial interest for the production of innovative materials in recent years. The general idea of this type of process route is that element-organic precursor molecules are constituted of structural units of the inorganic materials aimed for by this process, thus providing novel paths of controlling the atomic array, composition and microstructure of materials. Of special interest is the manufacture of amorphous or nano-crystalline covalently bounded inorganics on the basis of silicon, boron, carbon and nitrogen, revealing a unique potential of properties not known from conventionally prepared materials.

The preparation of high-quality SiC products requires a set of optimum synthesis conditions. Firstly, due to the carbothermal reduction of silica is a heterogeneous solid-state reaction; the mixing condition of two reactants greatly influences the properties of SiC products[7]. Recently, SiC materials with well defined morphology can be obtained by a $C/SiO_2$ artefact in a controlled atmosphere[8-9]. Furthermore, the gaseous environments during the whole reacting procedure also strongly affect the structures of the products[10].

In our previews work[11], it was found that SiC nanoparticles and nanofibers could be successfully fabricated via carbothermal reduction of mesoporous carbon-silica nanocomposites. Meanwhile, a small amount of carbon infiltrated into the mesopores can significantly change the partial pressure during the reaction and influence the microstructures of the SiC products. In Ref. 12, we infiltrated a preceramic precursor into the mesopores to form SiC nuclei first and then controlled the proportion of SiC nanoparticles and nanofibers.

In this paper, detailed characteristic changes of silicon carbide obtained from different precursors (mesoporous carbon-silica nanocomposites with different quantities of surface surfactant P123) by carbothermal reduction have been investigated. The objectives here are to understand influences of pores and gaseous environments on the formation of SiC by carbothermal reduction.

EXPERIMENTAL

Mesoporous silica/carbon composites were prepared by using our reported method. First, 5 g of $H_2O$, 3.28 g of ethanol (anhydrous, Aldrich) and 0.5 g of 1 M HCl (Merck) were mixed in a capped polypropylene bottle with a magnetic stirrer. To this solution, 4.1 g or 8.2 g of P123 ($EO_{20}PO_{70}EO_{20}$, MW5800, Sigma-Aldrich) was added under continuous agitation to obtain P123 solution. Then 10 g of tetraethoxysilane (TEOS) (99%, Sigma-Aldrich) and 3 g of furfuryl alcohol (FA, 99%, Aldrich) were added into the P123 solution. The resulting mixtures were rigorously stirred at room temperature for 3 h, followed by aging at room temperature for 4 days, and drying at 90 °C for 3 days. The black monoliths obtained were carbonized at 550 °C for 5 hours with flowing nitrogen gas, leading to mesoporous $C-SiO_2$ composites. To determine the $C/SiO_2$ ratios of $C-SiO_2$ composites, thermogravimetric analysis (TGA) was used to record the mass loss of the samples at a heating rate of 5 °C/min under flowing oxygen. Up to 700 °C all carbon was burned off. The $C/SiO_2$ ratio was then calculated by taking the total mass loss as the mass of carbon and the residual mass as the mass of $SiO_2$. The mesoporous $C-SiO_2$ composite had a $C/SiO_2$ molar ratio of 3.54/1 (Depending on the different quantity of P123, the samples were denoted as 3.54-CS-1P123 and 3.54-CS-2P123 respectively).

The $C-SiO_2$ composites were then transferred into a sealed tube furnace equipped with a vacuum pump. Before heating, the furnace was vacuumed to evacuate air for 10 min. To produce SiC, the $C-SiO_2$ composites were heated under argon atmosphere with a flowing rate of 250 ml/min and at a heating rate of 2 °C/min up to 1450 °C, and kept at this temperature for 5 hours. Then be cooled down to room temperature at a cooling rate of 2 °C/min. To investigate the effects of mesopores on the structure of SiC products, they were preheated at 1200 °C for different holding time before the final temperature.

Thermogravimetric analysis (TGA) (Perkin-Elmer, Pyris 1 thermogravimetric analyzer) was conducted at a heating rate of 5 °C /min to 700°C. X-ray diffraction (XRD) patterns were recorded on a Philips PW 1140/90 diffractometer with Cu Kα radiation at a scan rate of 2 °/min and a step size of 0.02°. Scanning electron microscopy (SEM) images were taken with a JSM-6300F microscope (JEOL). Transmission electron microscopy (TEM) images and selected-area electron diffraction (SEAD) were taken with a Philips CM20 microscope. The samples were degassed at 250 °C for 5 hours prior to the measurement. Nitrogen adsorption-desorption experiments were performed at 77 K by a Micromeritics ASAP 2020MC. The pore volume was estimated from the desorption branch of the isotherm at $P/P_0 = 0.98$ assuming complete pore saturation. The pore size distribution was calculated from the desorption branch of the isotherm by using Barrett-Joyner-Halenda (BJH) method.

RESULTS AND DISCUSSION

The nitrogen adsorption-desorption isotherms and pore size distributions of samples with a carbon/silica ratio of 3.54 with different quantities of P123 are shown in Fig. 1. The BET surface area and pore volume of these samples are summarized in Table 1. There was little shift between the adsorption-desorption hysteresis loops with different quantities of P123, which indicates that BET surface area, pore volume and pore size are increasing with the increase of P123.

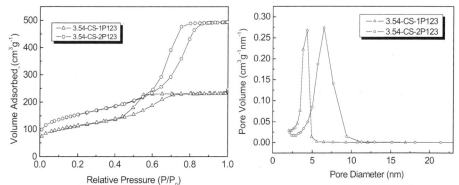

Fig. 1 Nitrogen results of mesoporous C-SiO$_2$ precursors with different quantities of P123

Table 1 Nitrogen adsorption-desorption results

| Samples | BET surface (m$^2$g$^{-1}$) | Peak pore size (nm) | Pore volume (cm$^3$g$^{-1}$) |
|---|---|---|---|
| 3.54-CS-1P123 | 393.41 | 4.32 | 0.36 |
| 3.54-CS-2P123 | 448.31 | 5.64 | 0.56 |

Fig. 2 shows the TEM images of 3.54-CS-1P123 and 3.54-CS-2P123, which exhibit mesoporous "wormlike" structures. With more P123, the pores became more and pore wall became thinner and weaker. SEM images (Fig. 3) and TEM images (Fig. 4) show the microstructure of SiC products obtained from 3.54-CS-1P123 and 3.54-CS-2P123 heated at 1450 °C for 5 hours. As can be seen, under the same heat treatment, particle size of nano SiC in sample 3.54-CS-2P123-1450/5h is a little bit larger than those in sample 3.54-CS-1P123-1450/5h.

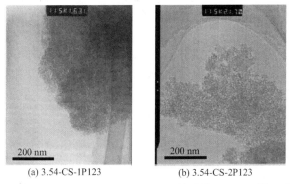

(a) 3.54-CS-1P123          (b) 3.54-CS-2P123

Fig. 2 TEM of mesoporous C-SiO$_2$ precursors with different quantities of P123

(a) 3.54-CS-1P123-1450/5h    (b) 3.54-CS-2P123-1450/5h

Fig. 3 SEM of SiC products obtained from mesoporous C-SiO₂ precursors with different quantities of P123

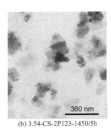

(a) 3.54-CS-1P123-1450/5h    (b) 3.54-CS-2P123-1450/5h

Fig. 4 TEM of SiC products obtained from mesoporous C-SiO₂ precursors with different quantities of P123

C-SiO$_2$ composites prepared without addition of P123 had a BET surface area of 51.1 m$^2$/g and a pore volume of 0.02 cm$^3$/g, which were considered that the interpenetrating C-SiO$_2$ structure was nearly dense[11]. It is expected that such an interpenetrating C-SiO$_2$ showed an extremely low permeability to gaseous species SiO, CO, and CO$_2$. In our work, 3.54-CS-1P123 was heated at 1200 °C for 2 hours or 15 hours. Table 2 show that 3.54-CS-1P123-1200/2h still maintain a mesoporous structure whereas the porosity of 3.54-CS-1P123-1200/15h is similar to the reference values of dense materials. Fig. 5 shows the XRD patterns of the product prepared by heating 3.54-CS-1P123 at different temperatures. A broad peak centered at about 23° suggested the presence of amorphous silica and carbon. The intensity of peak increased when there was longer time for preheating at 1200 °C. The SEM image (Fig. 6) shows that they were made up of nanoparticles.

Table 2 Nitrogen adsorption-desorption results

| samples | BET surface $(m^2g^{-1})$ | Peak pore size (nm) | Pore volume $(cm^3g^{-1})$ |
|---|---|---|---|
| 3.54-CS-1P123-1200/2h | 207.22 | 3.80 | 0.22 |
| 3.54-CS-1P123-1200/15h | 69.70 | 3.68 | 0.09 |

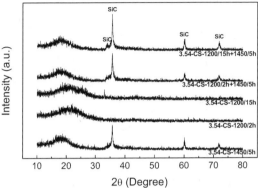

Fig. 5 XRD patterns of the products obtained from mesoporous C-SiO$_2$

(a) 3.54-CS-1P123-
1200/2h+1450/5h

(b) 3.54-CS-1P123-
1200/15h+1450/5h

Fig. 6 SEM images of the products obtained from mesoporous C-SiO$_2$

The traditional carbothermal reduction reaction between silica and carbon yields SiC through the following steps[12]:

$$SiO_2(s)+C(s) \rightarrow SiO(g)+CO(g) \qquad (1)$$
$$SiO_2(s)+CO(g) \rightarrow SiO(g)+CO_2(g) \qquad (2)$$
$$SiO(g)+C(s) \rightarrow SiC(s)+CO(g) \qquad (3)$$
$$SiO(g)+3CO(g) \rightarrow SiC(s)+2CO_2(g) \qquad (4)$$
$$CO_2(g)+C(s) \rightarrow 2CO(g) \qquad (5)$$

The overall reaction is:

$$SiO_2(s) + 3C(s) \rightarrow SiC(s) + 2CO(g) \qquad (6)$$

During the carbothermal reduction reaction, SiO$_2$ reacts with carbon leading to gaseous silicon monoxide (SiO). SiC is then produced by the reactions between SiO and C through eq. (3) or between SiO and CO through eq. (4). The equilibrium conditions at each step depend on temperature and partial pressure of SiO ($p_{SiO}$) and CO ($p_{CO}$).

In our case, the carbon content in the mesoporous pore walls gives higher partial pressure of SiO and CO. Following argon gas may carry away very little SiO and CO from the mesoporous C-SiO$_2$ samples because of small pore channels. SiC may nucleate through eq. (3) throughout the mesoporous precursors.

While the mesoporous C-SiO$_2$ nanocomposites had higher BET surface area and larger pore size, it significantly reduced the local concentrations of SiO and CO since there were more channels for the gas releasing. Locally concentrated SiO species readily react with carbon, resulting in a large number of SiC nuclei throughout the mesoporous C-SiO$_2$ nanocomposites, and larger nanoparticles formed during this process.

While the mesoporous C-SiO$_2$ nanocomposites were preheated at 1200 °C for 2 hours, the mesoporous structure still maintained, therefore, the products were similar with those without preheating, which were nanoparticles. While the mesoporous C-SiO$_2$ nanocomposites were preheated at 1200 °C for 15 hours, the pore size decreased dramatically. There are narrower porous channels for diffusion of the gaseous products, therefore, few SiC fibers formed.

CONCLUSIONS

We have shown that the pores of C-SiO$_2$ nanocomposites play a key role in the carbothermal reduction reaction, and the C/SiO$_2$ ratio and C-SiO$_2$ carbon structure are among other factors affecting the reaction kinetics and thus the nanostructures of resulting SiC. Mixed SiC nanofibers and nanoparticles were produced from mesoporous C-SiO$_2$ nanocomposites; the particle size of nanoparticles was tunable by changing the porosity of mesoporous C/SiO$_2$ nanocomposites and using preheating procedure. This is because the carbothermal reduction rate can be promoted by a mesoporous C/SiO$_2$ structure and a C-SiO$_2$ carbon interfacial structure, and mesoporous channels provide diffusion paths for the gaseous SiO, CO, and CO$_2$ products and confine SiC growth inside the mesoporous C-SiO$_2$ nanocomposites. The results presented here provide new insight into the formation of SiC nanostructures through carbothermal reduction. It should be possible to synthesize pure SiC nanostructures and mesoporous structures by carefully designing structures of their precursors.

ACKNOWLEDGEMENTS

The authors acknowledge the financial support for this work provided by the Australian Research Council and Monash University. Kun Wang gratefully acknowledges the assistance provided by Monash University through the Monash Graduate Scholarship (MGS) and Monash Fee Remission Scholarship (MFRS).

REFERENCES
[1]W. D. Kingery, D. R. Uhlmans, *Introduction to Ceramics*, New York, Wiley (1976).
[2]Acheson Industries Corporation, *Refractory Silicon Carbide*, New York, Springer (1977).
[3]A. W. Weimer, *Carbide, Nitride and Boride materials Synthesis and Processing*, Chapman & Hall (1997).
[4]S. Motojima, S. Gakei, H. Iwanaga, Preparation of Si$_3$N$_4$ whiskers by reaction of wheat husks with NH3, *Journal of Materials Science*, **30**, 3888-3892 (1995).
[5]J. Parmentier, J. Dentzer, C. Vix-Guterl, Formation of SiC via carbothermal reduction of a carbon-containing mesoporous MCM-48 silica phase: a new route to produce high surface area SiC, *Ceramics International*, **28**, 1-7 (2002).
[6]R. L.Beatty, Continuous silicon carbide whisker production, A. R. Company. US. Patent 4637924 (1987).
[7]X.K. Li, L. Liu, Y.X. Zhang, Synthesis of nanometer silicon carbide whiskers from binary carbonaceous silica aerogels, *Carbon*, **39**, 159-165 (2001).
[8]C. Vix-Guterl, I. Alix, P. Ehrburger, Synthesis of tubular silicon carbide from a carbon-silica

materials by using a reactive replica technique: Mechanism of formation of SiC, *Acta Materialia*, **52**, 1639-1651 (2004).

[9]G.W. Meng, Z. Cui, L.D. Zhang, Growth and characterization of nanostructured β-SiC via carbothermal reduction of $SiO_2$ xerogels containing carbon nanoparticles, *Journal of Crystal Growth*, **209**, 801-806 (2000).

[10]F.K. van Dijen, R. Metselaar, The chemistry of the carbothermal synthesis of β-SiC: reaction mechanism, reaction rate and grain growth, *J. Eur. Cera. Soc.*, **7**, 177-184 (1991).

[11]J.F. Yao, H.T. Wang, X.Y. Zhang, Role of pores in the carbothermal reduction of carbon-silica nanocomposites into silicon carbide nanostructures, *J. Phys. Chem. C*, **111**, 636-641 (2007).

[12]Kun Wang, Jianfeng Yao, Huanting Wang, Yi-Bing Cheng, Effect of seeding on formation of silicon carbide nanostructures from mesoporous silica-carbon nanocomposites, *Nanotechnology*, **19**, 175605 (8pp) (2008).

# SYNTHESIS AND CHARACTERIZATION OF MESOPOROUS NANOSTRUCTURED TiO$_2$-Al$_2$O$_3$ PHOTOCATALYTIC SYSTEM

M. L. García-Benjume[1], I. Espitia-Cabrera[2] and M. E. Contreras-García[1*].
[1] Instituto de Investigaciones Metalúrgicas, [2] Facultad de Ingeniería Química de la Universidad Michoacana de San Nicolás de Hidalgo. Edificio U, Ciudad Universitaria. Av. Francisco J. Mújica s/n. Colonia Felicitas del Río. Morelia, Michoacán, México.

## ABSTRACT

In the TiO$_2$-Al$_2$O$_3$ photocatalytic system, anatase (tetragonal titania phase) crystallizes at low temperatures from amorphous materials prepared by the simultaneous hydrolysis of titanium and aluminium alkoxides. In this work, nanostructured mesoporous TiO$_2$-Al$_2$O$_3$ photocatalyst, with 10, 20 and 30% Al$_2$O$_3$, were obtained by sol-gel hydrothermal synthesis and molecular self-assembly using Tween 20 as a surfactant. They were then calcined at 400 °C, at this temperature, only the anatase phase was obtained; specific surface areas up to 297 m$^2$/g were obtained with mean pore diameters from 6.6 to 9.2 nm presented type IV isotherms. The morphology and nanostructure of the obtained TiO$_2$-Al$_2$O$_3$ mesoporous photocatalyst were observed by SEM, TEM and HRTEM. Mean crystallite sizes from 4 to 5 nm were determined by TEM analysis.

## INTRODUCTION

A titanium oxide semiconductor has been increasingly proposed for different applications like bioceramics, photocatalysis, electronics etc., because a lot of new synthesis methods have been proposed in order to allow the design of structures at molecular level with unique textural and structural properties [1-4]. The traditional method of synthesis by sol-gel is complicated and requires special conditions. Using the molecular sieves method with organic surfactants, it is possible to obtain ordered mesoporous materials in friendly conditions, using ordinary equipment in conditions that are not complex. In this work we propose the synthesis of a mixed titania-alumina system using Tween 20 as a surfactant, in order to obtain a macro-mesoporous framework and to promote textural properties, because the aluminum oxide is easier to mesostructure than the transition metals oxides like titanium oxide [5-7] and also provides thermal and chemical stability, promoting textural properties [8].

In practical applications for separation membranes and catalysis mass transfer [9], the obtention of micro and mesoporous structural materials is of crucial importance, since the mobility of adsorbed molecules ultimately limits the rate of the overall process [10].

In the photocatalytic activity the TiO$_2$ phase is very important, particularly the anatase phase because it has a band gap below 3.2 eV., so it is possible to excite it with solar light, and it presents chemical inertness and good thermal stability [11]. X. He and D. Antonelli [12], described the catalytic activity of mesoporous titanium oxide compared with the nonporous anatase phase. They attributed the lower catalytic behavior to the amorphous nature of the walls of the mesoporous materials: the low degree of wall crystallinity and the high number of surface defects which lead to surface electron-hole recombination. The size of crystal and the crystallinity is then very important, so in this work the effect of the calcination temperature on the photocatalytic activity is studied.

EXPERIMENTAL PROCEDURE

Synthesis

The TiO$_2$ and TiO$_2$-Al$_2$O$_3$ nanostructured mesoporous photocatalyst were synthesized from titanium and aluminum butoxides (Aldrich, 99%) as precursors and an aqueous solution of surfactant Tween 20 using hydrothermal synthesis. The alkoxides were added drop by drop to this aqueous solution, and the mixture was stirred for 5 minutes. Then the sols were placed in an oven at a maximum temperature of 80 °C and were maintained at this temperature for 24 hours before being dried at 100 °C.

Samples with 10, 20, 30% mol of alumina concentration were prepared. Table I presents the description of the synthetized samples, indicating the composition of each one. The samples were labeled as TT (titania), TTA1 (alumina 10% mol), TTA2 (alumina 20% mol), TTA3 (alumina 30% mol) and AT (alumina). The obtained powders were calcined at 400 °C or 500 °C for 5 hours with a heating rate of 1 °C/minute. Samples labeled as TT400, TT500 refer to titania calcined at 400 °C and at 500 °C respectively.

Table I. Definition of the synthesis conditions of TiO$_2$, mixed oxides and Al$_2$O$_3$ samples.

| Sample | alumina composition % mol | Water (mol) | Surfactant (g) | Ti(OBu)$_4$ (mol) | Al ( OBu)$_4$ |
|--------|--------------------------|-------------|----------------|-------------------|----------------|
| TT | 0 | 1.5 | 3 | 0.01175 | 0 |
| TTA1 | 10 | 1.5 | 3 | 0.010575 | 0.001175 |
| TTA2 | 20 | 1.5 | 3 | 0.0094 | 0.00235 |
| TTA3 | 30 | 1.5 | 3 | 0.008225 | 0.003525 |
| AT | 40 | 1.5 | 3 | 0 | 0.01175 |

Characterization

The oxides and mixed oxides system were characterized by several methods. The obtained phases in the samples were determined by X-ray diffraction (XRD) in a Philips X'PERT model diffractometer, using Cu K$\alpha$ radiation $\lambda$= 1.54 Å, with 0.02°/seg steps. A scanning electron microscope (SEM), JEOL JSM-6400 model, was used to observe the morphology and the compositions. Transmission electron microscopy (TEM) was conducted with a field emission microscope, JEOL 2010. The Brunauer-Emmet-Teller (BET) surface areas of powders were calculated from nitrogen adsorption isotherms obtained with a Micrometrics ASAP 2010 model after the samples were degassed at 200 °C. The pore size distribution was determined by the Barrett-Joyner-Halenda (BJH) method.

For the photocatalytic test, the obtained photocatalyst powders were tested for methylene blue (MB) degradation. The batch reactor consisted of a reaction cell with a compressed air supply in a tightly closed compartment equipped with an ultraviolet light lamp (black light Ultravg l25 6286). The reaction was carried out at room temperature in aqueous solution with 10 mg/L of methylene blue. The ratio of photocatalyst was one gram of powder sample per liter of MB solution. The solution was previously magnetically stirred at 400 rpm for 30 minutes without the use of UV light in order to reach stability in the absorbance of MB on the photocatalyst surface [13]. After that, the solution was irradiated under UV light, magnetically stirred, and air was fed in simultaneously when the reaction system was started. Samples of the suspension were taken each hour during five hours of testing. The measurement of MB residual concentration in the solution was determined at 666nm for absorbance using a Perkin-Elmer UV-Vis Spectrophotometer Lambda-20 model. The concentration of MB after each experiment was determined quantitatively through the calibration graph constructed from standard solutions of MB at various concentrations.

RESULTS AND DISCUSSION
X-ray diffraction (XRD).
In order to elucidate the real crystallization temperatures of the synthetized titanium oxide, an X-ray diffraction study was carried out following the crystallization at different temperatures. In Figure 1, the diffractograms of the synthesized titania (TT) samples thermally treated at 400, 500, 600, 700 and 800 °C are presented. The main anatase peak corresponding to the (101) plane (diffraction card 21-1272) is wide and small indicating an incipient crystallization; this peak has greater intensity and definition until 600 °C, indicating the bigger size crystal formation. The anatase phase crystallization as a unique present phase is well defined until 700 °C. The diffractogram of the titania sample calcined at 800 °C also presented little peaks corresponding to the rutile phase. The TTA3400 sample diffractogram was included in the Figure 1 in order to show the broadening of the anatase peaks because the alumina presence. No peaks of alumina were obtained indicating the no crystallization of any alumina phase.

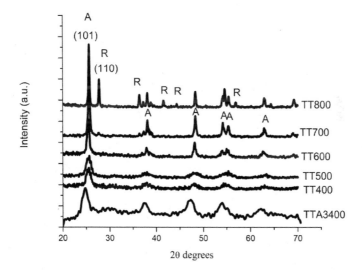

Figure 1. TT sample X-ray diffractograms calcined at different temperatures. Anatase phase diffraction peaks are labeled (A) and rutile phase diffraction peaks are labeled (R).
Figure 2 presents XRD patterns of the pure and mixed oxides with different compositions. The peaks labeled "A" correspond to anatase ICDD 21-1272. In samples TTA2 and TTA3, the peak of brukite phase is not evident, which indicates that alumina presence retards the brukite

formation. The pure titania sample presents the peak of brukite phase, labeled B, ICDD 29-1370 corresponding to (121) plane. Diffractograms of TTA2 and TTA3 samples show a diminished intensity of the peaks, which indicates that alumina addition, diminishes the crystallization.

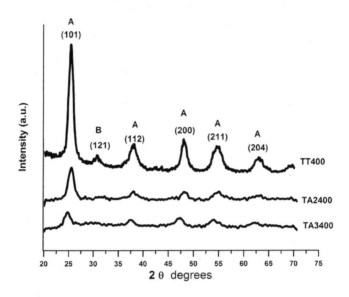

Figure 2. XRD patterns of the pure and mixed oxides with different compositions, Anatase phase diffraction peaks are labeled (A) and brukite phase diffraction peak is labeled (B).

Scanning electron microscopy (SEM).
Figures 3A and 3B show the SEM images of TT powders heat treated at 400 °C (TT400) at different magnifications. The TT400 sample is slightly rough, and is constituted by microspheres. The tween 20 used as a template directs the formation of porous titania with a bimodal meso-macroporous structure that has large ordered cavities connected to small channels formed by the decomposition of the surfactant micelles; the diameters of these pores are sub-micrometric. In Figure 3B, SEM observation of the obtained TT400 nano-sized powders shows homogeneous spherical consolidated particles 40-50 nm in size, forming porous walls with a very fine texture, which was also observed by TEM. This kind of structure is very useful as catalytic support for catalytic materials used in oxidative activity. This morphology also allows using them as membranes due to the permeability of the channels.

Figure 3. SEM morphologies of TT400 powers obtained: (A) 10,000 × and (B) at 40 000×.

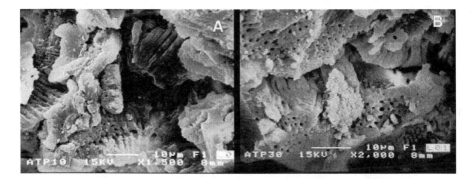

Figure 4. SEM micrographs: (A) TTA1400 and (B) TTA3400 samples.

The SEM morphologies of TTA1400 and TTA3400 samples are shown in Figures 4A and 4B. They show the powder structures formed by long ordered channels, X. Wang et al. [14] presented similar morphologies; the walls of the longitudinal channels are composed by microspheres as can be seen in figure 4A.

Nearly similar morphology was obtained for the alumina sample calcined at 400°C (AT400) as shown in Figure 5. The powders show a macroporous structure formed by large channels. All samples also have a mesostructure in the perpendicular long channel walls formed by nanometric microspheres.

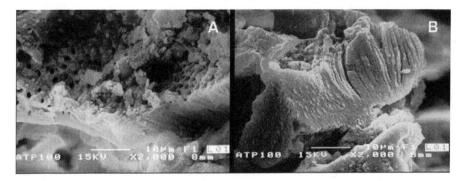

Figure 5. SEM micrograph of AT400 powders obtained

High resolution electron microscopy (HRTEM).

The HRTEM image from the TTA2400 sample of Figure 6A shows titania nanocrystals immersed in an amorphous matrix. In Figure 6B the size of the interplanar spaces of the titania nanocrystal is indicated and was measured as 3.52 nm, which corresponds to the (101) anatase main plane.

The anatase mean crystal size was analyzed by TEM observation and the results shown in Figure 7, the graphic indicates that the mean crystal sizes obtained are from 4 to 5 nm.

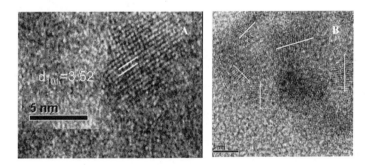

Figure 6. HRTEM images of $TiO_2$ (A) anatase nanocrystal indicating the (101) interplanar distance and (B) different sizes of $TiO_2$ nanocrystals.

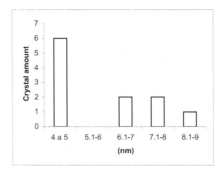

Figure 7. Anatase nanocrystal mean sizes obtained from HRTEM images of the TTA2400 sample.

The small mean size of the crystals can be attributed to the incipient crystallization obtained at calcination temperature of 400 °C.

Brunauer-Emmet-Teller (BET).
The Nitrogen adsorption-desorption isotherms and BJH pore size distribution were performed on the adsorption branch of mesoporous titania-alumina mixed oxide phases after calcinations, synthesized using Tween 20. Figure 8 shows the Nitrogen adsorption-desorption isotherm and pore size distribution of mesoporous TTA3400. This isotherm is representative of all samples, because in all cases we found a type IV isotherm representative for mesoporous solids with H2 hysteresis loop, indicating the existence of the mesoporous texture. In Figure 8 the specific surface area of sample TA3400 is 297 m$^2$/g and the average pore diameter is 9.2 nm. The specific surface area data of the samples calculated by the BET equation are shown in Table II. The pore size distribution was calculated using the BJH method, assuming a cylindrical pore model and using the adsorption branch of the isotherms. The values obtained for all the studied samples were unimodal with pore sizes ranging from 50 to 150 Å. The values are shown in table II. It is important to note that the specific surface area diminishes as the alumina content decreases. These results are in accordance with those obtained by J. M. Dominguez [15] in this system without use of surfactant.

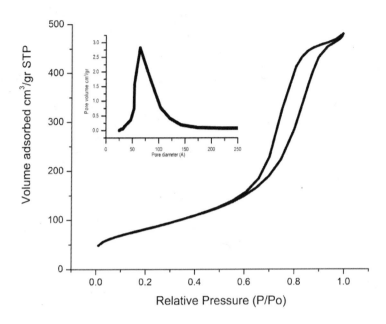

Figure 8. N$_2$ absorption-desortion Type IV isotherm of TTA3 sample. Inset: Pore size distribution.

Table II. Textural properties of the TiO$_2$, TiO$_2$-Al$_2$O$_3$ and Al$_2$O$_3$ nanostructured mesoporous photocatalyst

| Sample | BET surface area (m$^2$/gr) | Pore Diameter (BJH adsorption) (nm) |
|--------|-----------------------------|-------------------------------------|
| TT400 | 136 | 8.5 |
| TTA10400 | 146 | 8.6 |
| TTA2400 | 235 | 8.8 |
| TTA3400 | 297 | 9.2 |
| AT400 | 413 | 12.1 |

Photocatalytic activity

Today´s environmental concerns demand degradation of pigments, and titanium oxide is an effective photocatalyst for this application. Therefore the photocatalytic activities of the obtained samples were investigated by detecting the decomposition of the MB aqueous solution under UV light irradiation. Figure 9 shows the MB concentration reached before UV light exposition and the degradation behavior of the different samples. The samples TT400 and TTA1400 showed similar parabolic behavior and also reached similar degradation percentages, of 48 and 47% respectively, after 5 hours' testing. Sample TTA2400 presented a lineal continuous behavior, reaching 45% degradation, which was better than TTA3400, which reached 40%, after 5 hours' testing. The percentage degradation is higher for samples with low alumina content. However, best degradation observed, of 63%, was for sample AT2500. This means that when the samples are treated at higher temperatures, 500 °C in this case, the photocatalytic activity increases. This is certainly due to the better crystallization of the titania in this sample.

Therefore, the effect of the addition of alumina was evidently beneficial on the textural properties and morphology of the samples. However the effect on the photocatalytic activities is not very clear. As was determined by XRD, the alumina retards the titania crystallization; otherwise, the titania's photocatalytic behavior has been attributed to the crystallization degree, and the effect of the titania crystallite size on the photocatalytic behavior has been explained by different authors [12, 16]. From the obtained results, it appears that a low alumina addition amount helps the photocatalyst function by adding textural conditions that favor the light transmission and the transport of reactants liquid through the macro channels and the mesoporous walls. But a higher amount of alumina addition than 20% mole can represent a physical obstacle for the titania-light interaction during the photocatalytic process. The highest photocatalytic behavior (63% of MB degradation), obtained for the ATT2500 sample, demonstrated that a better crystallization of titania is more beneficial for the photocatalytic behavior than the highly specific surface area of 297 m²/g obtained for this sample.

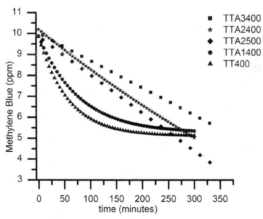

Figure 9. Methylene Blue concentration versus time.

CONCLUSION

Alumina additions to the titania photocatalyst allowed to easily obtain highly ordered, macro-mesoporous TiO₂-Al₂O₃ with 0, 10, 20, 30 and 100% Al₂O₃ content structures have been synthesized using tween 20 as the surfactant. The anatase phase was formed and the alumina addition effect was evaluated, as a retarding agent of the crystallization process. The titania BET surface area has been improved with alumina addition. The obtained macro-mesoporous structure with nanocrystalls in the walls of $TiO_2$ and $TiO_2$-$Al_2O_3$ presented good photocatalytic activity and presented thermal stability at 400°C. The mesoporous walls favored the methylene blue adsorption. The effect of alumina addition on the photocatalytic activity was mainly presented in the MB degradation behavior, changing it from parabolic, with pure and 10% mole alumina compositions to lineal and continuous behavior with alumina additions greater than 20%. These characteristics make these powders a highly effective photocatalyst because the interdifusion resistance is minimized. The obtained sponge type structure with a macroporous network has high capacity to absorb light and with some surface modifications these powders are also promissory for other novelty optoelectronic applications.

ACKNOWLEDGEMENTS
The authors express their thanks to Jose Reyes Gasga, Gabriel Solana Espinosa and Ana María Nuñez Gaytán, for their contribution to the techniques used in the research, and to Francisco Solorio and Victoria Luke for their technical support of this research, which was performed through CONACYT (Project No. 60888) and CIC-UMSNH.

REFERENCES
[1] L. Hakim, J. McCormic, G. Zhan and A. Weimer, Surface Modification of Titania Nanoparticles Using Ultrathin Ceramic Films, J. Am. Ceram. Soc., 89, 3070-3075 (2006).
[2] T. Ishikawa, Photocatalytic Fiber with Gradient Surface Structure Produced from a Polycarbosilane and Its Applications, J. Appl. Ceram. Technol., 1, 49-55 (2004).
[3] S. Baldassari, S. Komarneni, E. Mariani, C. Villa, Rapid Microwave-Hydrothermal Synthesis of Anatasa to Form of Titanium Dioxide, J. Am. Ceram. Soc., 88, 168-171 (2005).
[4] F. Schüt and Wolfang Schmidt, Microporous and Mesoporous materials, Adv. Eng. Mat., 4, 269-279 (2002).
[5] J. Cejka, P. Kooyman, L. Veselá, J. Rathousky and A. Zukal, High-Temperature Trasnformations of Organized Mesoporous Alumina, Phys. Chem. Chem. Phys., 4, 4823-4829 (2002).
[6] J. Cejka, Organized Mesoporous Alumina: Synthesis, Structure and Potential in Catalysis, Appl. Catal. A, 254, 327-338 (2003).
[7] X. Zhao, Y. Yue, Y. Zhang, W. Hua, and Z. Gao, Mesoporous Alumina Molecular Sieves: Characterization and Catalytic Activity in Hydrolysis of Carbon Disulfide, Catal. Lett. 89, 41-47 (2003).
[8] N. Yao, G. Xiong, Y. Zhang, M. He and W. Yang, Preparation of Novel Uniform Mesoporous Alumina Catalyst by the Sol-Gel Method, Catal. Today, 68, 97-109 (2001).
[9] J. Adler, Ceramic Diesel Particulate Filters, Int. J. Appl. Ceram. Technol., 2, 429-439 (2005).
[10] J. Kárguer and D. Fraude, Mass Transfer in Micro-and Mesoporous Materials, Chem. Eng. Thechnol., 25, 769-778 (2002).
[11] M. Kaneko, I. Okura, Photocatalysis Science and Technology, eds. Kodansha Springer, Japan, 1-2 (2002).
[12] X. He and D. Antonelli, Recent Advances in Synthesis and Applicatiosn of Transitions Metal Containing Mesoporous Molecular Sieves, Angew. Chem. Int. Ed. 41, 214-229 (2002).

[13]Q. Sheng, S. Yuan, J. Zhang and F. Chen, Synthesis of mesoporous titania with high photocatalytic activity by nanocrystalline particle assembly, Microporous and Mesoporous Materials, 87, 177–184 (2006).

[14]X. Wang, J. C. Yu, Chunman Ho, Y. Hou and X. Fu., Photocatalytic Activity of Hierarchically Macro/ Mesoporous Titania, Langmuir, 21, 2552-2559 (2005).

[15]J. M. Domínguez, J. L. Hernandez and G. Sandoval, Surface and Catalytic Properties of $Al_2O_3$-$ZrO_2$ Solid Solutions Prepared by sol-gel Methods, Appl. Catal. A, 197, 119-130 (2000).

[16]S. Yang and L. Gao, Preparation of Titanium Dioxide Nanocrystallite with High Photocatalytic Activities, J. Am. Ceram. Soc., 88, 968-970 (2005)

# MONODISPERSED ULTRAFINE ZEOLITE CRYSTAL PARTICLES BY MICROWAVE HYDROTHERMAL SYNTHESIS

Michael Z. Hu[1]*, Lubna Khatri[2], Michael T. Harris[3]

[1]Oak Ridge National Laboratory, Oak Ridge, TN 37831-6181
[2]Department of Chemical Engineering, University of Maryland, College Park, MD 20742
[3]Department of Chemical Engineering, Purdue University, Lafayette, IN 47905
(*Corresponding Author, *E-mail address*: HUM1@ORNL.GOV, *Phone*: 865-574-8782)

ABSTRACT

Monodispersed ultrafine crystal particles of zeolite (Silicalite-1) have been synthesized in batch reactor vessels by microwave irradiation heating of aqueous basic silicate precursor solutions with tetra propyl ammonium hydroxide as the templating molecule. The effects of major process parameters (such as synthesis temperature, microwave heating rate, volume ratio (i.e., the volume of the initial synthesis solution relative to the total volume of the reactor vessel), and synthesis time on the zeolite particle characteristics are studied using a computer-controlled microwave reactor system that allows real-time monitoring and control of reaction medium temperature. The changes in the morphology, size and crystal structure of the particles are investigated using scanning electron microscope, dynamic light scattering, X-ray diffraction, and BET surface analysis. We have found that the synthesis temperature, volume ratio, and heating rate play a significant role in controlling the particle size, uniformity, and morphology. Microwave processing has generated morphologies of zeolite particles (i.e., uniform block-shaped particles that contain mixed gel-nanocrystallites and agglomerated crystal particles) that are not typically made by conventional hydrothermal processes. At higher synthesis temperatures and lower volume ratios, irregular block-shaped particles were produced, whereas increasing the volume ratio promoted the formation of monodispersed single-crystal particles with uniform shape. Our results clearly demonstrate that faster microwave heating is advantageous to enhance the zeolite crystallization kinetics and produces larger-size crystal particles in shorter time. In addition, zeolite crystallization mechanisms, depending on the microwave heating rate, are also discussed.

## 1. INTRODUCTION

Microwave processing of materials is a technology that can provide the material processor with a powerful tool to synthesize materials that may not be amenable to conventional means of processing and/or to improve the performance characteristics of existing products [Clark and Sutton 1996; Rao et al. 1999]. Microwave irradiation is more efficient for transferring thermal energy to a volume of material than conventional thermal processing, which transfers heat through the surfaces of the material by convection, conduction, and radiation. Microwave dielectric heating is based on molecular friction or inductive heating through the conducting properties of the synthesis mixture The oscillating electromagnetic field, generated by microwaves interacts with the dielectric properties of materials, leading to rotation of molecular dipoles and subsequent energy dissipation as heat from internal resistance to that rotation [Bonaccorsi and Proverbio 2003]. Such heating is more volumetric (efficient energy transfer through penetrating radiation, electromagnetic wave resonance or relaxation), can be very rapid and selective (through differential absorption of materials at microscopic level, such as specific

91

energy dissipation via microwave energization of the hydroxylated surface or associated water molecules in the boundary layer, and formation of the high potential of the active water molecules). Table I briefly summarizes the main features of microwave heating. Rapid heating can reduce reaction time, increase production yield, and induce more simultaneous nucleation and growth. Microwave processing could improve heating uniformity and homogeneity as well as controllability through out the reaction vessel, and thus, improve product performance, provide higher energy efficiency, and reduce overall costs.

Table I. Microwave-Induced Heating Versus Conventional Heating

| PARAMETER | MICROWAVE HEATING | CONVENTIONAL HEATING |
|---|---|---|
| Heating Mechanism | Volumetric | Depends on Heat Transfer |
| Heating Rate | Very High | Limited |
| Heating Uniformity | Uniform | Non-uniform |
| Efficiency | High | Low |

Hydrothermal synthesis, a common process for producing ultrafine and nanosized particles, often utilizes conventional heating, introducing potential problems of non-uniform heat distribution in the reactive medium. In contrast, microwave volumetric heating is more uniform in liquid solution and thus can promote more homogeneous reaction conditions and homogeneous nucleation. Such conditions are desirable for producing uniform-sized particles, rapid synthesis, favorable formation of small crystallites, facile morphology control, and so on. Therefore, microwave processing of liquid solutions (particularly for aqueous solutions, called "microwave hydrothermal" processing) has been investigated as a promising approach for synthesis of ceramic/metal powders and nanoparticles [Komarneni et al. 2004; Liang et al. 2002; Zhu et al. 2002; Bellon et al. 2001; Bondioli et al. 2001; Li and Wei 1998; Ma et al. 1997; Moon et al. 1995]. However, due to the complexity of microwave interactions with materials or molecules in solutions, it is necessary to fully understand the reaction-dependent nucleation and crystal growth during microwave synthesis. General issues for microwave hydrothermal processing of zeolites include the following, to mention a few: (1) Insufficient understanding of nucleation and growth mechanisms/kinetics, (2) Lack of detailed studies on effect of process parameters (heating rate, starting synthesis solution volume, and aging time prior to synthesis, temperature, reaction time, etc.), (3) Rate of degradation of (organic) templating molecules with applied microwave radiation and its effect on nanostructure of the final product, and (4) incomplete knowledge of the effects of solution vaporization at higher synthesis temperatures on particle nucleation and growth and final product characteristics?

Zeolites are microporous crystalline aluminosilicates with a framework of corner-sharing $TO_4$ tetrahedra (in which the T-sites are occupied either by silicon or aluminum). Zeolites have rather complex but precisely repetitive, ordered, atomic network with submicroscopic channels or pores that are typically 3 to 10 Å in size [Ghobarkar et al. 1999; Breck 1974; Smith 1976]. These cavities and pores are uniform in size within a specific zeolite material. Zeolite ZSM-5 can be synthesized by using variety of templating species such as quarternary ammonium compounds (TEA-Br, TEA-OH, and TPA-OH), amines and organic compounds [Somani et al. 2003]. One special ZSM-5 type of zeolite is Silicalite-1, which is aluminum-free. Small crystal particles of Silicalite-1 can be synthesized by solution-based hydrothermal process, as we have reported in the past [Khatri et al. 2003]. In addition to the traditional use of zeolites as catalysts and adsorbents, zeolite-based advanced inorganic materials (for electronic devices and membranes applications) require fabrication strategies that provide control over the morphology and size of the forming building-block crystal. The early stages of the formation of crystal particles from solution can be decisive for the properties of the final crystal solid. Microwave heating could be utilized as a means to affect the nucleation event and thus later stage particle growth. Therefore, it is necessary to know how microwave heating makes the difference from the conventional heating and affects the zeolite production in terms of their particle size, uniformity, morphology, and crystallinity.

Literature Review of Microwave Hydrothermal Synthesis of Zeolite Particles

Special applications of microwave heating technology to the hydrothermal synthesis processes are rather new and gaining more importance since the early nineties, especially, for the syntheses of (i) nanoporous materials such as zeolites (zeolite A, zeolite Y, TS-2, and ZSM-5 etc.) [Serrano et al. 2004; Park et al. 2004; Somani et al. 2003; Phiriyawirut et al. 2003a, 2003b; Romero et al. 2004], (ii) mesoporous materials such as MCM-41 and SBA-15 etc. [Wu and Bein 1996, Park et al. 1998], and (iii) $AlPO_4$-type microporous materials [Braun et al. 1998]. The general belief is that microwave rapid heating (1) causes simultaneous, abundant nucleation and fast dissolution of gel, and (2) enhances the crystallization rate in a very short time and reduce the overall synthesis time, leading to small particle size with narrow particle size distribution and high phase purity [Somani et al. 2003; Sathupunya et al. 2002; Kim et al. 2004]. In addition, microwave is considered a very efficient tool to control the morphology of zeolite or other microporous materials (such as TS-1, MCM-41, SBA-16, and AFI-type molecular sieves) [Jhung et al. 2004].

One initial work on microwave hydrothermal crystallization of zeolites was reported by Chu et al. (1990, 1988), which has pointed out the possibility that microwave-induced heating increases the rate of crystallization (or productivity) and directs the crystallization mechanism. The rapid crystallization rates permit one to obtain smaller crystals within a shorter period than is possible with conventional hydrothermal processes and with less risk of contamination. The crystallinity of the synthesized ZSM-5 zeolite samples, using seeded growth method, was used to compare the hydrothermal and microwave synthesis methods. However, that study does not provide any information on the morphology and other characteristics of the synthesized particles. Also, the microwave synthesis experiments were controlled by constant power, not by temperature or vapor pressure, thus, the actual temperature and vapor pressure inside the reaction vessel are not known during the synthesis.

Microwave heating was used in the unseeded preparation of Zeolite Y and the seeded preparation of monoclinic ZSM-5 [Arafat et al. 1993]. Very short crystallization times using microwaves were ascribed to relatively fast dissolution of the gel upon microwave irradiation as compared to conventional heating systems. It was also recognized that the organic templating molecule ($TPA^+$) is subject to Hoffman degradation during the synthesis, and microwave radiation might accelerate this degradation. Macroscopically, microwave heating was found to produce small crystals of "elongated prismatic shape" while oven heating leads to "cube-shaped" crystals. But the influence of the microwave heating on the nanostructure inside a crystal and characteristics of the final ZSM-5 product were still not clear.

Through previous studies by Slangen et al. (1997), it is known that during microwave heating both the ions and the dipoles in the synthesis mixture (the water molecules) are used to generate heat. Thoroughly mixed and aged solution systems yielded Na zeolites, like NaA, in a very short time. Still, the possibility of ion-mediated self-assembly of particles through controlled microwave radiation needs to be verified.

The mechanisms for microwave enhancement of crystallization rate were investigated by Cundy et al. (1998) and attributed to differential microwave heating effects due to the heterogeneity of the dielectric medium (i.e., the reaction mixture which contains colloidal or particular materials). They believed that local superheating could result from a number of energy-loss mechanisms: (i) dipolar polarization losses varying with local composition, (ii) interfacial (Maxwell-Wagner) polarization losses, and (iii) conduction losses associated with clusters or arrays of ions. Thus, another reason for microwave enhanced crystallization rate may derive from an acceleration of reagent digestion and silicate equilibration processes to produce a medium which is more homogeneous on a molecular level. An even more significant effect may be occurring at crystal surfaces, where microwave energization of the hydroxylated surface or of associated water molecules in the boundary layer may be linked to specific energy dissipation through modes (ii) and (iii) above.

Recently, Somani et al. (2003) reported that a microwave-hydrothermal heating technique was superior to conventional hydrothermal heating in reducing the time period for crystallization of ZSM-5 zeolite, while the similar synthesized product (in terms of framework rigidity or structural stability) was maintained. They found that crystallinity of the samples increased with the crystallization period and the hydrothermal crystallization rate was further enhanced by the microwave treatment of precursor hydrogel after room-temperature aging. They obtained crystallization kinetic curve showing two distinct stages: (i) an induction period, when the nuclei are formed; and (ii) a crystal growth period, when the nuclei grow into crystals.

Besides particles and powders, microwave techniques have also been studied for the purpose of synthesizing films of zeolites (such as zeolite A, NaY, and oriented Silicalite-1) [Cheng et al. 2003, 2002; Baek et al. 2001; Katsuki et al. 2001; Xu et al. 2001, 2000; Han et al. 1999]. For example, Han et al. (1999) used a power-controlled microwave heating scheme and demonstrated that microwave heating greatly reduced the reaction time for zeolite A crystallization from precursor hydrogels. More uniform and smaller crystals were generally found in zeolite membranes prepared by microwave heating.

The present paper will specifically focus on Silicalite-1 and report the results of process parametric studies on microwave-hydrothermal synthesis of zeolite solid particles from precursor solutions. The objectives of this work are (i) to study the effects of process parameters on the characteristics of synthesized zeolite, (ii) to understand the nucleation, growth, and crystallization mechanisms in microwave-hydrothermal synthesis, and (iii) to clarify any differences between microwave-assisted and conventional hydrothermal processing relative to the conventional hydrothermal synthesis, microwave-assisted process is more flexible in tailoring crystallization mechanisms and in controlling size and morphology and properties of zeolite solid particles.

## 2. MATERIALS AND METHODS

Synthetic zeolites are generally prepared by processing an aqueous solution of the desired oxides and other required compounds of the crystallization reaction under heat and pressure. In this work, crystals of TPA-Silicalite-1 (i.e., TPA occluded Silicalite-1) were formed according to the Exxon patent [Verduijn 1992]. The schematic procedure for templated hydrothermal synthesis is described in Figure 1. The silicalite synthesis solution was prepared in a capped Teflon container by dissolving sodium hydroxide pellets (99.99 %, Aldrich) in a tetra propyl ammonium hydroxide (TPAOH or TPA$^+$) solution (1M in water, Aldrich). The mixture was then heated and kept at 80°C while fumed silica (99.98%, Aldrich) was dissolved under vigorous stirring to obtain a clear solution. Filtered (0.25 μm filter) deionized water was then added so that the final molar ratio of the synthesis solution was 10 SiO$_2$ : 3 TPAOH : 1.05 NaOH : 140 H$_2$O (or mass ratio of 20 g SiO$_2$ : 100 mL TPAOH : 1.4 g NaOH : 3.2 g H$_2$O). The solution was cooled and aged at room temperature for 3 h.

The clear solution was then transferred to a Teflon-lined, stainless-steel autoclave or a microwave reactor vessel and heated to the specified synthesis temperature for a period of time. Conventional heating was conducted in a cylindrical shaped, Teflon lined, stainless steel bomb placed in a convection oven. Microwave heating was accomplished in a special computer-controlled microwave oven processing system (QLAB 6000, Questron Technologies Corp., Missisauga, Ontario, Canada) (Figure 2). The microwave oven features power up to 1000 W (at a frequency of 2450 MHz), constant temperature, pressure, or power-level control options, and offers controllable heating rate. The microwave reactor vessels (very high pressure VHP) are made of composite engineering materials (ULTEM) internally lined with TFM Teflon bottom cup and PTFE Teflon top lid. The VHP vessel (with 80-mL internal volume) can be operated up to 230 °C and 625 psig.

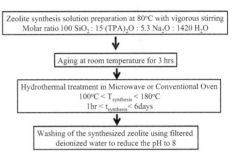

Figure 1. Templated hydrothermal synthesis of Silicalite-1 crystal particles.

For each synthesis run, the closed reactor vessel containing predetermined sample

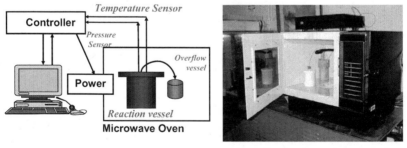

Figure 2. Computer-controlled microwave hydrothermal processing system. (Left) Schematic illustration of various components. (Right) Photo picture.

volume ($R_v$) of aged solution was placed in the microwave cavity, connected with temperature and pressure sensors, and heated at a pre-specified ramp time ($t_R$) to a target synthesis temperature ($T_{synth}$) for a certain period of time ($t_{synth}$) typically 1 h after reaching the target $T_{synth}$). The microwave power was automatically controlled with the temperature feedback probe (coated with Teflon) in the VHP vessel to maintain a constant temperature synthesis medium. At the end of each run, reaction medium must cool down ($< 60^{\circ}C$) before opening the vessel lid. The resulting solids were retrieved by centrifugation of reaction mixture at 20,000 rpm using a high- speed ultracentrifuge, and thoroughly washed with several batches of filtered deionized water until the pH of the wash water was around 8.

All samples were prepared under similar conditions using the procedure described above. Volume ratio ($R_v$) is defined as the ratio of the initial synthesis solution volume to the total internal volume ($\sim$ 80 mL) of the synthesis vessel. The ramp time $t_R$, a parameter for controlling microwave heating rate, is defined as the time to heat the synthesis solution to reach the target synthesis temperature $T_{synth}$.

The size, morphology and microstructure of the collected solid particles were investigated using a scanning electron microscope (SEM) (Jeol JSM-T220A). The effective hydrodynamic diameter of zeolite particles was measured by a custom designed dynamic light scattering (DLS) spectrophotometer (Hu et al. 1998), which is suitable for measuring particle sizes, typically ranging from 5nm to 1μm. Powder X-ray diffraction (XRD) was used to characterize the crystal structure and crystallinity of samples. Data acquisition was performed using DMS-NT software (Scintag Inc., Cupertino, CA) and data analysis was undertaken using Jade software (Materials Data Inc., Livermore, CA). For specific surface area measurements, a nitrogen gas adsorption-based BET analyzer (Gemini III2375) by Micromeritics (Norcross, GA) was used. Samples were calcined in a furnace at $500^{\circ}C$ for 12 h before doing the BET analyses.

## 3. RESULTS AND DISCUSSION

Our study has shown that conventional hydrothermal synthesis of zeolite (Silicalite-1) crystal particles takes long time (usually several days) [Khatri et al. 2003]. One problem in previous studies has been the long heating and cooling times compared to the total synthesis time in a regular Teflon-lined stainless steel autoclave: heating up the synthesis mixture in a conventional preheated oven to $\sim 180^{\circ}C$ took more than 1 hour; it is possible that solid gel or crystal nucleation could occur during the heating process. For this reason, it was not possible to obtain a clear picture of the parametric effect of synthesis temperature on the nucleation rate and crystal growth rate [Joegler et al. 1997]. Microwave heating takes advantage of homogeneous heating throughout a reaction vessel, offering us a better tool to enable and to study homogeneous nucleation processes. The results reported below shows that microwave heating shorten the solid gel particle formation time and crystallization time. This is consistent with previous findings reported by Wu and Bein (1996) and others. In the case of zeolite synthesis, molecular nature and dynamics variation of microwave interaction with different reactants in the liquid (such as water, dissolved silica species, and organic templating molecule TPA$^+$) could change the reaction mechanism and offer possible means of controlling a resulting zeolite nanostructure. We believe that the microwave heating mechanism accelerates the hydrolysis and condensation reactions of the silicate network and thus the solid gel nucleation and crystallization processes.

Below we report some original results of the process parametric studies in microwave-hydrothermal synthesis of Silicalite-1 crystals.

Induction/Aging Period

The crystallization of zeotype materials is frequently constrained by limitations at the nucleation stage, so it is common practice to age reaction solution mixtures or add nucleants (or seeds) to them [Cundy et al. 1998]. Aging and nuclei addition offer two beneficial effects: (1) reduction of the induction period preceding the detection of crystalline product and (2) promotion of a dominant crystalline phase that is usually similar to the seeding material. These two effects lead to shortened overall synthesis time and improved product purity. Cundy et al. (1998) suggested that the rearrangement of the synthesis mixture to yield crystal nuclei is the bottleneck in a microwave synthesis. Insufficient aging and inadequate stirring cause the formation of impurities in the final product and low crystal yield. The limiting factor was identified as a lack of nuclei formation in the synthesis mixture during the short microwave heating time [Bonaccorsi and Proverbio 2003]. Cundy et al (2003) recently demonstrated that room-temperature aging of the precursor solutions, prior to crystallization step, helps generating "proto-nuclei" that mature into viable nuclei and grow into crystals upon subsequent heating.

(A) $T_{synthesis} = 150°C$

(B) $T_{synthesis} = 180°C$

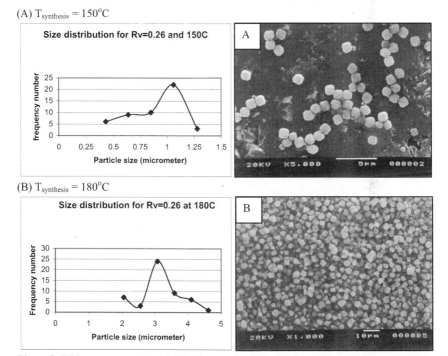

Figure 3. DLS measurements and SEM images for Silicalite-1 particles synthesized at $R_v = 0.26$ and $t_R = 1$ min. (A) $T_{synthesis} = 150°C$ (DLS size = 1001.6 nm, SEM size = 881 nm), (B) $T_{synthesis} = 180°C$ (DLS size = ~3476.5 nm, SEM size = 3090 nm).

For the Silicalite-1 system, we also found that the time length of room temperature aging affects the nucleation and crystallization kinetics of the later-stage hydrothermal processing. Generally, longer aging time leads to faster gellation and zeolite crystallization. Through this study, we focus on the parametric effects during microwave synthesis and fixed the aging time at 3h. For all experiments, the synthesis time ($t_{synth}$) was fixed as 1 h after ramping up to the target synthesis temperature.

(A) $T_{synthesis} = 150^\circ C$

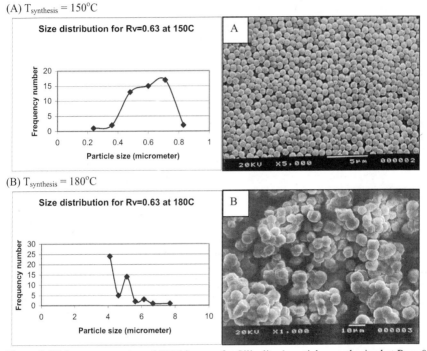

(B) $T_{synthesis} = 180^\circ C$

Figure 4. DLS measurements and SEM images for Silicalite-1 particles synthesized at $R_v = 0.63$ and $t_R = 1$ min. (A) $T_{synthesis} = 150^\circ C$ (DLS size = 693.4 nm, SEM size = 598 nm), (B) $T_{synthesis} = 180^\circ C$ (DLS size = ~3006.4 nm, SEM size = 4800 nm).

Changes in Particle Morphology Due to Synthesis Temperature

A set of experiments was carried out to compare the effect of synthesis temperature on the particle size and morphology. The samples were quickly heated to the synthesis temperature ($T_{synthesis} = 150$ and $180^\circ C$) with 1-min ramp time, at two initial volume ratio conditions ($R_v = 0.26$ and $0.63$). Results based on SEM images and DLS measurements are shown in Figure 3 and 4, suggesting that the synthesis temperature greatly affects the particle morphology, size and microstructure of the Silicalite-1 particles. At the lower temperature ($150^\circ C$), individually dispersed single-crystal particles were always obtained (Figure 3A and 4A), but smaller crystals correspond to the higher $R_v$ conditions. However, higher temperature ($180^\circ C$) leads to large irregular block-shaped particles (Figure 3B) or the visible agglomeration of crystal particles (Figure 4B). Note that the temperature effect depends on the $R_v$ value. Under lower $R_v$ condition (0.26), higher temperature tends to produce irregular block-shaped particles, each of which contains extremely small nanocrystallites and some condensed gel components. XRD identified the existence of Silicalite-1 crystalline phase in the sample, but the crystallinity of the particles has not been determined. Under higher $R_v$ condition (0.63), higher temperature leads to the visible agglomeration of large crystal particles.

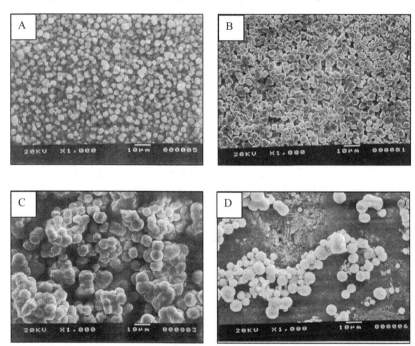

Figure 5. SEM images for Silicalite-1 particles synthesized at various volume ratio ($R_v$) conditions ($T_{synthesis}$ = 180°C and $t_R$ = 1 min). (A) $R_v$ = 0.26 (DLS size = 3476.5 nm, SEM size = 3090 nm), (B) $R_v$ = 0.38 (DLS size=2354.8 nm, SEM= 2420nm), (C) $R_v$= 0.63 (DLS size=3006.4 nm, SEM= 4840nm), (D) $R_v$= 0.88 (DLS size=1864.7 nm, SEM= 2400nm).

These results can be explained by the possible liquid-vapor reaction phenomena. At the lower temperature (150°C), a minor amount of vapor is generated in the closed reactor vessel, and liquid phase co-exists in equilibrium. The concentration of reactant (i.e., dissolved silica species) in the existing liquid phase depends on the $R_v$ value. For example, the actual reactant concentration after evaporation at lower $R_v$ (0.26) is higher than the concentration at higher $R_v$ conditions (0.63). It is known that higher concentration should lead to faster particle growth kinetics and larger final particles. This concentration effect explains very well the crystal particle size difference between Figure 3A and 4A.

Figure 6. SEM images of Silicalite-1 particles obtained at various microwave heating rate ($t_R$) conditions with $T_{synthesis}$ = 180°C and $R_v$ = 0.63. (A) $t_R$ =1 min (SEM size = 4940 nm), (B) $t_R$ =30 min (SEM= 770nm), (C) $t_R$ =60 min (SEM size = 380nm).

However, at higher temperature (such as 180°C), evaporation of liquid phase into vapor becomes very significant and the liquid phase could be converted into vapor phase due to microwave heating. The reactant concentration in the liquid phase increases drastically with heating time. Amorphous gel solid nucleation and crystal growth in the liquid phase only have limited time to occur and will eventually stop due to the complete vaporization, which result in semi-dried (i.e., vapor-soaked or extremely concentrated) solid/gel sludge at the bottom of the reactor vessel. Continued condensation and crystallization could still occur in the vapor-soaked solid phase under microwave heating. In our current reactor set up, the time point for complete evaporation can not be observed or determined. However, we do know that the evaporation rate at lower $R_v$ should be much faster than the one at higher $R_v$. This means that the time for liquid-phase nucleation and crystal growth at lower $R_v$ is much shorter and only gel solids nucleate and some amount of small nanocrystallites grow in the short existence of liquid phase. The irregular, block-shaped particles shown in Figure 3B indicate such semi-drying of gel-nanocrystal mixed sludge. On the other hand, higher $R_v$ conditions allow a longer time for liquid-phase nucleation and crystallization, long enough to grow into large crystals before the more complete liquid evaporation into vapor. Vaporization appeared to have caused the fusion of some crystals. This discussion explains results shown in Figure 4B (i.e., the agglomerated large crystal particles). Above a certain $R_v$, liquid phase will exist in equilibrium with the vapor phase, and thus, nucleation and crystallization occur in the liquid phase and agglomeration could be inhibited.

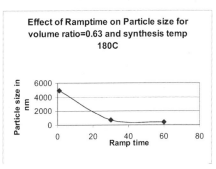

Figure 7. Dynamic light scattering (DLS) measurements of particles sizes.

So far, a few works have been reported using microwave irradiation to control crystal morphology of other zeolite molecular sieves [Jhung et al. 2004], but there are no reports on Silicalite-1. Because many emerging applications of porous materials require precise control of crystal morphology, strategies to control the crystal shape and size are necessary for special applications. This study shows the possibility of microwave-controlled production of various morphology, size, and microstructure of Silicalite-1 solids by varying process conditions. The particles synthesized at a lower synthesis temperature of 150°C and larger volume ratio are more monodispersed and comparatively smaller than the particles synthesized at a higher temperature of 180°C. At the higher temperature, agglomerated and irregular block-shaped zeolite particles were formed. This can be due to rapid heating to high temperature, as the ramp time is only 1 min. Vaporization during early synthesis might result in fused nuclei, which grow together to form large agglomerated clumps.

Figure 8. SEM images of Silicalite-1 particles obtained at various microwave heating rate ($t_R$) conditions with $T_{synthesis}$ = 180°C and $R_v$ = 0.26. (A) $t_R$ =1 min (SEM size = 3150 nm), (B) $t_R$ =30 min (SEM= 900nm), (C) $t_R$ =60 min (SEM size = 290nm).

The strategy in microwave processing is to find a material that is polarizable and whose dipoles can reorient rapidly in response to the changing electric field strength. However if these materials have poor thermal conductivity, heat does not rapidly dissipate into the surrounding regions of the material when a region in the solid becomes hot. This difficulty is compounded, because the dielectric loss frequently increases dramatically as the temperatures. Thus the hot region becomes even hotter, sometimes resulting in local melting. These "hot spots" are a major difficulty and might be responsible for the fused nuclei and coalescence of the particles in the semi-solid sludge (Figure 4B).

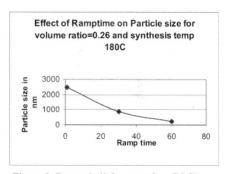

Figure 9. Dynamic light scattering (DLS) measurements of particles sizes.

At higher temperature and lower volume ratio, vaporization is much faster and therefore the hot spot effects might be enhanced microscopically in the "dried" gel-nanocrystallite solid phase (Figure 3B).

Volume Ratio ($R_v$) Affects the Formation of Dispersed Nanoparticles of Silicalite-1

Some discussions related to the $R_v$ effect have been made above. The second set of experiments focused on understanding $R_v$ effects at the higher synthesis temperature of 180°C. More detailed results on various $R_v$ values are shown in Figure 5. It is observed that rapid heating causes the formation of irregular, block-shaped, large particles at lower $R_v$ values (from 0.26 to 0.63), whereas dispersed small crystal particles are obtained at higher $R_v$ (0.88). X-ray diffraction data (not shown here) confirmed that the agglomerated or block-shaped particle samples also contained amorphous gel, which cannot be removed by washing. We believe that during synthesis, liquid solution must have been evaporated into vapor and such vaporization directly affected the solid particle characteristics (solid block size, crystal aggregation state, and crystallinity). The vaporization affected the pathway of crystallization processes, different from those occurring in the bulk liquid phase. It was previously established that the enthalpy of vaporization is the same under microwave and conventional heating, but the rate of evaporation is dependent on microwave intensity [Saillard et al. 1995]. Higher intensity of rapid microwave heating tends to induce rapid evaporation, particularly true when the $R_v$ value is low. Microwave provides much higher heating rate ($t_R$ as short as 1 min) to target temperature than the conventional heating that usually takes ~ 1-2 h to heat up to target temperature. This may explain why we have not observed such vaporization-induced agglomeration process in the conventional hydrothermal process. It is also noted that the temperatures of both vapor and liquid at the liquid/vapor interface strongly depend on the experimental conditions and particularly on the absorbed microwave power.

Figure 10. SEM images of Silicalite-1 particles obtained at various microwave heating rate ($t_R$) conditions with $T_{synthesis}$ = 180°C and $R_v$ = 0.38. (A) $t_R$ =1 min (SEM size = 2480 nm), (B) $t_R$ =30 min (SEM= 850 nm), (C) $t_R$ =60 min (SEM size = 240 nm).

Monodispersed Nanoparticles Synthesized With Controlled Heating Rate

Now that it is evident that the sample volume ratio ($R_v$) can change particle morphology, it is necessary to study the particle morphology when the heating rate (here, indicated by the heating-up ramp time $t_R$) is changed. The third set of experiments was carried out at a fixed synthesis temperature ($T_{synthesis}$) of 180°C, with the volume ratio ($R_v$) of 0.63.

SEM images of obtained Silicalite-1 particles, shown in Figure 6, indicate that by increasing ramp time (or decreasing heating rate) to reach the synthesis temperature, the particle size decreases while the shape uniformity and monodispersity of the silicalite-1 particles improve. In other words, increasing the microwave heating rate has enhanced the crystal growth rate and thus the final crystal particle size. Note that this observation is conflicting to the intuitive reasoning that faster conventional heating could cause rapid nucleation of larger number of nuclei, traditionally leading to smaller particle size in the fixed volume of reaction solution. Here, DLS measurements (Figure 7) also indicate that faster microwave heating has led to the production of larger crystal particles, in agreement with SEM observations (Figure 6). This result clearly demonstrates that faster microwave heating is advantageous to enhance the zeolite crystallization kinetics. At a temperature of 180°C and low ramp time of 1 minute, the gel converts into solid, and therefore, results in irregular shaped particles due to limited dissolution and restructuring of the species, originally existing in liquid solution phase. The SEM of such a sample is shown in Figure 6A, where the sample is agglomerated and form large sized clumps. These irregularly shaped agglomerates contain large Silicalite-1 crystals, as confirmed by room-temperature powder X-ray diffraction. At longer ramp time of 30 minutes (Figure 6B) and 60 minutes (Figure 6C), the gel goes through dissolution and reconstruction in the solution phase, therefore, a controlled crystallization of Silicalite-1 nanoparticles takes place, leading to dispersed, uniform-sized crystals.

Similarly, lower volume ratios of 0.26 and 0.38 were also studied to prepare samples with various ramp times ($t_R$ = 1min, 30min, or 60 min). First the samples were prepared using a volume ratio of 0.26 at the three different ramp times. The SEM images of these samples are given in Figure 8 and corresponding DLS size measurements are shown in Figure 9. Results for samples made at volume ratio of 0.38 are shown in Figure 10 and 11. In summary, regardless of the volume ratios, faster heating rates clearly enhanced the crystal particle growth rate, although it also induced the rapid nucleation of large number of initial nuclei, which usually lead to the production of smaller crystals in a conventional heating process.

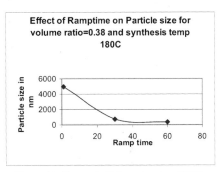

Figure 11. Dynamic light scattering (DLS) measurements of particles sizes.

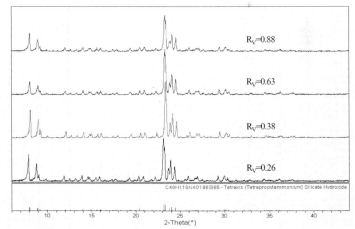

Figure 12. X-ray diffraction analyses of samples obtained at various volume ratios. Conditions: $T_{synthesis}$ = 180°C, $t_R$ = 1 min, $t_{synthesis}$ = 1 h, calcined at 500°C for 12 h.

Effect of Volume Ratio on Surface Area

XRD analyses (Figure 12) confirmed that as-prepared samples at all volume ratios are particles containing TPA-Silicalite-1 zeolite crystals. This indicates that crystallization in vapor-soaked gel (at low $R_v$ ratio) does occur under microwave heating, although its kinetics and particle-particle interaction may be different from those in bulk solution phase. After calcination at 500°C for 12 h, they still maintained the Silicalite-1 crystal structure. The samples were further analyzed using BET nitrogen adsorption to measure the surface area of the particles made at various volume ratio conditions (Figure 13). After activation (i.e., calcination at 500°C for 12 h), samples show specific surface areas in the range from 330-380 $m^2$/g. Such difference is considered significant. It is interesting to notice that the specific surface area depends on the volume ratio, indicating the microstructure difference in these particle samples. Higher specific surface area corresponds to sample obtained at low $R_v$ (0.26), indicating a possibility of the existence of smaller nanocrystals and pores in the relatively large, block-shaped particles (Figure 5A). Samples containing larger crystal particle sizes (Figure 5C and 5D) correspond to smaller surface areas. While we consider that crystal size might affect the specific surface area, we can not exclude the possibility that intra-crystal cage structure variation might be also responsible for the significant difference of surface areas.

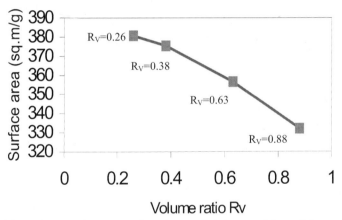

Figure 13. The surface area decreases as the volume ratio is increased for particles synthesized at 180°C using microwave-induced heating. Conditions: $T_{synthesis}$ = 180°C, $t_R$ = 1 min, $t_{synthesis}$ = 1 h, calcined at 500°C for 12 h.

Comparing Conventional Hydrothermal with Microwave-Induced Synthesis

There is a significant difference when a sample is synthesized using different heating methods, as shown by SEM images. Figure 14 shows monodispersed 200-nm particles made by using a conventional heating method (180°C, 1h). In contrast, Figure 5C shows agglomerated clumps of large crystal particles (~ 5 microns), synthesized by microwave-induced heating using a 1-min ramp time ($t_R$) and 59-min synthesis time ($t_{synthesis}$). Both samples were prepared using 3hr aged synthesis solution of the same chemical composition at $R_v$ = 0.63, $T_{synthesis}$ = 180°C, and total synthesis time of 1hr. Conventional hydrothermal conditions produced smaller individual crystal particles, indicating a slower crystal growth rate than microwave-induced process. Note that the heating rate in the conventional oven is not controllably uniform during long warming time. In contrast, microwave heating offers a controllable, more uniform heating rate in a short ramp time. With relatively slow heating rate ($t_R$ = 1 h ramp time, equivalent to the total heating up time in conventional heating above), beautiful monodispersed crystal particles (Figure 6C) could be produced by microwave processing.

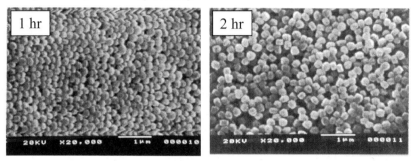

Figure 14. Silicalite-1 crystal particles (~200 and 250 nm) obtained by conventional hydrothermal synthesis at 180°C for 1 h and 2 h (Note: this time include heating up time).

It is known that nucleation rate is affected by the heating rate, which can be enhanced via microwave heating. Higher nucleation rate leads to smaller particle sizes. On the other hand, crystal growth is fairly temperature-sensitive. When conventional hydrothermal processes are utilized to obtain relatively small crystals, it has intrinsically slow heating rate and is necessary to reduce the temperature, and thus, concomitantly increase the time for complete crystallization to occur [Chu et al., 1990]. Therefore, conventional hydrothermal processes are poor in productivity to produce small crystals. In contrast, microwave heating processes allow rapid nucleation of large number of gel particles and production of smaller crystals within a shorter period than is possible with the conventional heating processes. Indeed, this work indicates that microwave hydrothermal processing offers more flexibility than the conventional heating in tailoring particle size and uniformity as well as generating new morphologies that conventional heating could never achieve.

Mechanisms of Solid Crystal Formation
Synthesis of Silicalite-1 crystal particles starts with the aged, but visually clear liquid solution. In general, two popular mechnisms of zeolite crystal formation in the literature include:
(i) the "solution-mediated transport" mechnism, in which amorphous gel is dissolved to provide reactants for nucleation that ultimately changes into crystal, and
(ii) the "solid phase transformation" mechanism, in which amorphous gel is assumed to be converted into crystal.
Our recent studies [Khatri et al., 2003] have shown that formation of amorphous gel spheres occurs as a transition step prior to the crystallization of zeolite (silicalite-1), i.e., solution → gel → crystal. Crystal nucleation may start from the center of a gel particle and crystal growth appears to consume surrounding gel materials. A similar process was also reported for the TS-2

zeolite synthesis, proceding from solid gel to solid crystal [Serrano et al. 2004], which is very different from the mechanism proposed by Somani et al. (2003).

In the case of ZSM-5 studied by Somani et al. (2003), the solution-mediated transport mechanism was suggested to be occurring as the exposure to microwave enhanced the dissolution of the gel contents. Microwave irradiation was believed to destroy the hydrogen bridges of the water molecules by ion oscillation and water dipole rotation, resulting in so-called active water. The lone pair of OH group of active water molecules has greater potential to dissolve the gel constituents than normal water [Arafat et al. 1993, Jansen et al. 1992], thus reducing the induction period and making the crystallization process faster.

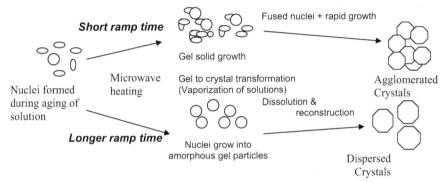

Figure 15. Schematic of Silicalite-1 crystal particle formation mechanisms.

In our case with the Silicalite-1, we believe that the zeolite crystallization involves more complicated processes due to the possibility of vaporization of liquid solutions, depending on the synthesis conditions such as heating rate ($t_R$) and volume ratio ($R_v$). It can be reasoned that the controllable rapid rate of microwave heating clearly accelerates the initial-stage hydrolytic condensation reactions (at the molecular level) and thus the initial-stage gel solid nucleation and growth. On the other hand, microwave heating also tends to dissolve the gel solid with a faster kinetic rate than conventional heating. Both condensation and dissolution rate enhancement lead to faster rearragement kinetics of molecular silicate species into zeolite crystal or onto the surface of an existing crystal. Our data clearly show that faster heating rate (i.e., shorter ramp time) enhances the crystal growth rate (Figure 6 and 7). Such an overall enhancement may be attributed to the advantageous microwave effect on each individual event, such as gel solid nucleation, gel dissolution, crystal nucleation from rearrangement of molecular species, or crystallization by deposition/rearrangement of molecular species on an existing crystal surface. In addition, results obtained by Koegler et al. (1997) support the similar conclusion that microwave heating effectively converts the gel precursor into zeolite crystals. After the gel precursors are completely consumed, the crystal could still grow by consuming the residual dissolved silicate species from solution. We believe that there is a competition between the two mechanisms for zeolite crystallization. When the heating rate is relatively low (like in the conventional hydrothermal synthesis conditions [Khatri et al. 2003]), "solid phase transformation" mechanism may be dominating. However, when the heating rate is fast enough (like in the very rapid heating conditions by microwave [Somani et al. 2003]), the transient state of solid gel exist very short time and fast dissolution of the gel dominates. Therefore, the "solution mediated transport" mechanism could be responsible for the zeolite crystalization.

In this work, we have explored a new phenomenon of two-phase reactions due to possible vaporization of liquid solutions under rapid heating. We suggest the following possible mechanism, as illustrated in Figure 15, which could explain the agglomeration process occurring under conditions of low volume ratio and short ramp time as well as the process that produced monodispersed crystals. The reason behind agglomeration behavior is possibly due to formation of fused crystal nuclei in the thick gel phase during rapid microwave heating, coupled with liquid vaporization. However, under slow heating conditions, liquid solution phase is maintained and nucleation and growth phenomena occur dominantly in the solution. Uniform microwave heating and Ostwald ripening helps the formation of monosized, dispersed crystals. The schematic of Figure 15 illustrates the various morphologies of zeolite particles we have observed with SEM. However, the issues on formation of the initial location of crystal nucleation (i.e., in the core of gel particles, at the gel-liquid interface, or in the liquid solution) and the effects of microwave heating on structure templating of TPAOH molecules need further studies to clarify.

## 4. CONCLUSION

Microwave hydrothermal processing is a comparatively new and fascinating method to produce high-quality zeolite particles. In this study, microwave processing was successfully used to produce Silicalite-1 materials with various controllable sizes, morphologies, and crystallinity. The synthesis process parameters (such as synthesis temperature, heating rate, and volume ratio), significantly affect the product characteristics. It is shown that, regardless of the volume ratio used, the crystal particles size increased with decreasing ramp time, indicating that rapid microwave heating enhances the crystal growth rate. Agglomeration decreased with increasing ramp time, and near-monodispersed crystal particles can be obtained by maintaining liquid-phase reactions at larger volume ratios and longer ramp times. In contrast to conventional heating processes, microwave heating can generate various morphologies of zeolite solid particles. A mechanistic model has been proposed to interpret the particle sizes and morphologies obtained under various conditions.

## ACKNOWLEDGMENTS

This work was sponsored by the Division of Materials Science (KC 02 03 01 0), Office of Science, the U.S. Department of Energy. Research is also sponsored in part by the National Science Foundation (DMR-9700860). Thanks are given to E. A. Payzant for his assistance in XRD analyses. L. Khatri is supported by an appointment to the Professional Internship Program (PIP) administered jointly by the Oak Ridge Institute of Science and Education and the Oak Ridge National Laboratory, managed by UT-Battelle, LLC, for the U.S. Dept. of Energy under contract DE-AC05-00OR22725.

## REFERENCES

A. Arafat, J. C. Jansen, A. R. Ebaid, H. van Bekkum, "Microwave preparation of zeolite Y and ZSM-5", Zeolites 13, 162-165 (1993).

D. Baek, U. Y. Hwang, K. S. Lee, Y. Shul, K. K. Koo, "Formation of zeolite A film on metal substrates by microwave heating," J. Ind. Eng. Chem. 7, 241-249 (2001).

K. Bellon, D. Chaumont, D. Stuerga, "Flash synthesis of zirconia nanoparticles by microwave forced hydrolysis," J. Mater. Res. 16, 2619-2622 (2001).

F. Bondioli, A. M. Ferrari, C. Leonelli, C. Siligardi, G. C. Pellacani, "Microwave-hydrothermal synthesis of nanocrystalline zirconia powders," J. Am. Ceram. Soc. 84, 2728-2730 (2001).

L. Bonaccorsi, E. Proverbio, "Microwave assisted crystallization of zeolite A from dense gels," J. Crystal Growth 247, 555-562 (2003).

I. Braun, G. Schulz-Ekloff, D. Wohrle, W. Lautenschlager, "Synthesis of AlPO$_4$-5 in a microwave-heated, continuous-flow, high-pressure tube reactor," Microporous and Mesoporous Mater. 23, 79-81 (1998).

D. W. Breck, Zeolite Molecular Sieves, Wiley, New York, 1974.

Z. L. Cheng, Z. S. Chao, W. P. Fang, H. L. Wan, "Synthesis regularity of NaA zeolite membrane in microwave field," J. Inorg. Mater. 18, 1306-1312 (2003).

Z. L. Cheng, Z. S. Chao, H. L. Wan, "Research of a type zeolites as well as zeolite membrane by microwave technique heating," Chinese J. of Inor. Chem. 18(5), 528-532 (2002).

P.-C. Chu, F. G. Dwyer, J. C. Vartuli, "Crystallization method employing microwave radiation," EP358827A1, March 21, 1990.

P.-C. Chu, F. G. Dwyer, J. C. Vartuli, " Crystallization method employing microwave radiation," US Patent 4,778,666, Oct. 18, 1988.

D. E. Clark, W. H. Sutton, "Microwave processing of materials," Annu. Rev. Mater. Sci. 26, 299-331 (1996).

C. S. Cundy, J. O. Forrest, R. J. Plaisted, "Some observations on the preparation and properties of colloidal silicates. Part I. synthesis of colloidal silicate-1 and titanosilicate-1 (TS-1)," Micro. Meso. Mater. 66, 143-156 (2003).

C. S. Cundy, R. J. Plaisted, J. P. Zhao, "Remarkable synergy between microwave heating and the addition of seed crystals in zeolite synthesis," Chem. Commun., 1465-1466 (1998).

H. Ghobarkar, O. Schaf, U. Guth, "Zeolites — from kitchen to space," Prog. Solid St. Chem. 27, 29-73 (1999).

Y. Han, H. Ma, S. Qiu, F.-S. Xiao, "Preparation of zeolite A membranes by microwave heating," Microporous and Mesoporous Mater. 30, 321-326 (1999).

M. Z.-C. Hu, M. T. Harris, C. H. Byers, "Nucleation and growth for synthesis of nanometric zirconia particles by forced hydrolysis," J. Colloid Inter. Sci. 198, 87-99 (1998).

J. C. Jansen, A. Arafat, A. K. Barakat, H. van Bekkum, in: M. L. Occelli, H. E. Robson (Eds.), Synthesis of Microporous Mater. Vol. 1, Van Nostrand Reinhold, New York, p. 507, 1992.

S. H. Jhung, J. S. Chang, Y. K. Hwang, S.-E. Park, "Crystal morphology control of AFI type molecular sieves with microwave irradiation," J. Mater. Chem. 14, 280-285 (2004).

H. Katsuki, S. Furuta, S. Komarneni, "Microwave versus conventional-hydrothermal synthesis of NaY zeolite," J. Porous Mater. 8, 5-12 (2001).

L. Khatri, M. Z. Hu, M. T. Harris, "Nucleation and growth mechanism of Silicalite-1 nanocrystal during molecularly templated hydrothermal synthesis," Ceram. Trans. 137, 3-21 (2003).

D. S. Kim, J.-S. Chang, J.-S. Hwang, S.-E. Park, J. M. Kim, "Synthesis of zeolite beta in fluoride media under microwave irradiation," Micro. Meso. Mater. 68, 77-82 (2004).

J. H. Koegler, A. Arafat, H. van Bekkum, J. C. Jansen, "Synthesis of films of oriented silicalite-1 crystals using microwave heating," Progress in Zeolite and Microporous Materials (Editors: H. Chon, S.-K. Ihm, and Y. S. Uh), Studies in Surf. Sci. Cat. 105, 2163-2170 (1997).

S. Komarneni, H. Katsuki, D. Li, A. S. Bhalla, "Microwave-polyol process for metal nanophase," J. Phys.: Condens. Matter 16, S1305-S1312 (2004).

Q. Li, Y. Wei, "Study on preparing monodispersed hematite nanopartilces by microwave-induced hydrolysis of ferric salts solution," Mater. Res. Bull. 33, 779-782 (1998).

J. H. Liang, Z. X. Deng, X. Jiang, F. L. Li, Y. Do. Li, "Photoluminescence of tetragonal ZrO2 nanoparticles synthesized by microwave irradiation," Inorg. Chem. 41, 3602-3604 (2002).

Y. Ma, E. Vileno, S. L. Suib, P. K. Dutta, "Synthesis of tetragonal BaTiO3 by microwave heating and conventional heating," Chem. Mater. 9, 3023-3031 (1997).

Y. T. Moon, D. K. Kim, C. H. Kim, "Preparation of monodispersed ZrO2 by the microwave heating of zirconyl chloride solutions," J. Am. Ceram. Soc. 78, 1103-1106 (1995).

S. E. Park, J. S. Chang, Y. K. Hwang, D. S. Kim, S. H. Jhung, J. S. Hwang, "Supramolecular interactions and morphology control in microwave synthesis of nanoporous materials," Catalysis Surveys From Asia 8, 91-110 (2004).

S. E. Park, D. S. Kim, J.-S. Chang, W. Y. Kim, "Synthesis of MCM-41 using microwave heating with ethylene glycol," Catalysis Today 44, 301-308 (1998).

P. Phiriyawirut, R. Magaraphan, A. M. Jamieson, S. Wongkasemjit, "Morphology study of MFI zeolite synthesized directly from silatrane and alumatrane via the sol-gel process and microwave heating," Microporous and Mesoporous Mater. 64, 83-93 (2003a).

P. Phiriyawirut, R. Magaraphan, A. M. Jamieson, S. Wongkasemjit, "MFI zeolite synthesis directly from silatrane via sol-gel process and microwave technique," Mater. Sci. Eng. A361, 147-154 (2003b).

K. J. Rao, B. Vaidhyanathan, M. Ganguli, P. A. Ramakrishnan, "Synthesis of inorganic solids using microwaves," Chem. Mater. 11, 882-895 (1999).

M. D. Romero, J. M. Gomez, G. Ovejero, A. Rodriguez, "Synthesis of LSX zeolite by microwave heating," Mater. Res. Bull. 39, 389-400 (2004).

R. Saillard, M. Poux, J. Berlan, M. Audhuy-Peaudecerf, "Microwave heating of organic solvents: thermal effects and field modeling," Tetrahedron 51, 4033-4042 (1995).

M. Sathupunya, E. Gulari, S. Wongkasemjit, "ANA and GIS zeolite synthesis directly from alumatrane and silatrane by sol-gel process and microwave technique," J. Eur. Ceram. Soc. 22, 2305-2314 (2002).

D. P. Serrano, M. A. Uguina, R. Sanz, E. Castillo, A. Rodriguez, P. Sanchez, "Synthesis and crystallization mechanism of zeolite TS-2 by microwave and conventional heating," Micro. Meso. Mater. 69, 197-208 (2004).

P. M. Slangen, J. C. Jansen, H. van Bekkum, "Induction heating: a novel tool for zeolite synthesis," Zeolites 18, 63-66 (1997).

J. V. Smith, "Origin and Structure of Zeolites," In J. A. Rabo (ed.), Zeolite Chemistry and Catalysis. American Chemical Society Monograph 171, Washington D.C., 1976, pp. 3-79.

O. G. Somani, A. L. Choudhari, B. S. Rao, S. P. Mirajkar, "Enhancement of crystallization rate by microwave radiation synthesis of ZSM-5," Mater. Chem. Phys. 82, 538-545 (2003).

J. P. Verduijn, "Process for producing substantially binder-free zeolite," WO 92/12928, August 6, 1992.

C.-G. Wu, T. Bein, "Microwave synthesis of molecular sieve MCM-41," Chem. Commun., 925-926 (1996).

X. H. Xu, W. H. Yang, J. Liu, L. W. Lin, "Synthesis of NaA zeolite membrane by microwave heating," Separation and Purification Technology 25 (1-3), 241-249 (2001).

X.-C. Xu, W.-S. Yang, J. Liu, L.-W Lin, "Synthesis of a high-permeance NaA zeolite membrane by microwave heating," Adv. Mater. 12, 195-198 (2000).

J.-J. Zhu, J.-M. Zhu, X.-H. Liao, J.-L. Fang, Miao-Gao Zhou, H.-Y. Chen, "Rapid synthesis of nanocrystalline SnO2 powders by microwave heating method," Mater. Lett. 53, 12-19 (2002).

# THE STRUCTURE OF NANOPARTICULATE AGGREGATES OF TITANIA AS A FUNCTION OF SHEAR

M. Jitianu[1], C. Rohn[2], R. A. Haber[1]

[1]Rutgers, The State University of New Jersey, 607 Taylor Rd, Piscataway, NJ 08854, USA
[2]Malvern Instruments Inc., 117 Flanders Road, Westborough, MA 01581-1042, USA

ABSTRACT

The understanding of rheological properties of suspensions of titania particles is of great importance for a lot of industrial processes. The rheological behavior of such suspensions employing two kinds of $TiO_2$ from a sulfate process has been investigated. The goal was to understand the influence of synthesis parameters in the aggregation behavior of these powders. Aggregate strength was investigated via oscillatory stress rheometry. Specifically, G' and G'' were measured as a function of the applied oscillatory stress. Oscillatory rheometry indicated a strong variation as a function of the sulfate level of the particles in the viscoelastic yield strengths. To assess different degrees of aggregation of titania powders in suspensions, a modified freeze-drying technique combined with FE-SEM imaging has been developed to investigate the degree of aggregation at different stages of rheological measurements. The results will be discussed taking into account powder characteristics, including electrokinetic properties.

## INTRODUCTION

Titanium dioxide ($TiO_2$), especially anatase, is extensively used in catalysis as filters, adsorbents and catalyst supports[1] and in pigments and composites industry, as well[2]. Moreover, nanostructured titania has received growing attention recently because of photocatalytic and photovoltaic application[3]. Agglomeration characteristics of $TiO_2$ particles are of great importance in each industrial application since dispersed particles are desired for pigments or composites[4], but agglomerated ones are desired when preparing catalyst pellets to facilitate reactant/product flow[5].

A number of techniques have been employed to measure agglomerate size, such as applied or electrical mobility techniques, small angle scattering[6] and more recently, monitoring real-time agglomerate growth[7]. To assess agglomeration in post-synthesis processes, agglomerates can be ruptured by mechanical stirring, low energy agitation[8]. When studying aggregation of particles in suspensions, for dilute suspensions, aggregate sizes can be measured by optical techniques, such as light scattering[9], or sedimentation[10], but such techniques are not suitable for high solids concentration, usually associated with extrusion compositions with applications in catalyst industry. Only few recent studies investigated a correlation between aggregation and overall suspension rheology, particularly in concentrated suspensions[11]. If particle-particle interaction in a suspension lead to aggregation, the effects on the viscosity would be very significant since, for example, the aggregates are bigger than the individual particles, therefore more resistant to flow[12]. Moreover, the aggregates immobilize some of the liquid phase, leading to an increase in the apparent solids volume fraction which results in a

111

higher viscosity at low shear rates along with a non-Newtonian shear thinning behavior of the suspension[12].

Generally, suspensions encountered in industry contain particles of various sizes and most such slurries have average particle sizes higher than the colloidal dimensions. Nevertheless, there are usually enough sub-micron particles to ensure that electrostatic interparticle forces to be significant, so mutual aggregation of the components is an important factor influencing the flow properties of such a suspension.

The $TiO_2$ employed in this present work has particles between 0.17-0.7 $\mu$m, and has been synthesized via sulfate process[13], where modification of particular parameters resulted in powders varying in residual sulfate level. The agglomerate structure of these powders will be assessed via rheological behavior of concentrated aqueous suspensions. It has been shown that the rheological behavior at slow flow rates is extremely dependent upon agglomeration[13]. In suspensions of high solids concentration, rheological profiles obtained for dynamic stress rheometry can be used to compare the strength of bonding between agglomerates for different $TiO_2$ powders with different levels of residual sulfate[14].

Russel[15] established that colloidal forces control the rheological properties of fine particle suspensions, and these forces are significant for particle sizes less than 1$\mu$m. The $TiO_2$ particles used in this work have a sub-micron size distribution; hence the colloidal forces will be significant. Hunter[16] defined that the two main colloidal forces are the van der Waals attractive forces originating from fluctuating dipoles and the electrostatic repulsive forces due to the presence of charges on the particles and a dielectric medium. Generally, for a given system, the van der Waals would be essentially constant. The electrostatic forces will vary with the surface charge density of the suspended particles. Residual sulfate amounts on $TiO_2$ particles investigated in this present work are expected to affect electrostatic forces in suspensions created with different sulfate-level $TiO_2$.

Studies on rheology of the $TiO_2$ in aqueous systems in the absence of organic dispersants have received less attention[17]. Therefore, the purpose of this work is to assess agglomeration in concentrated aqueous suspensions containing titania powders which will be further used in catalysis applications. The $TiO_2$ powders contain different residual-sulfate levels and study will evaluate suspension rheology along with FE-SEM imaging of freeze dried suspensions at different moments of rheological investigation.

EXPERIMENTAL

$TiO_2$ samples used in this work are commercially available, synthesized from ilmenite ore via sulfate process and consist of anatase with two different levels of residual sulfate, expressed as $SO_3$ % (wt%). One of the samples contains 0.4% residual sulfate and labeled as "low sulfate titania", the sulfate content of the other one being 3.4%, and it will be named "high sulfate titania".

A Bohlin rotational rheometer (Malvern Instruments Ltd. UK) was employed in the rheological studies. The vane-cup system was used for all rheological tests, which have been conducted at T=25°C, temperature controlled by a Peltier Couette system. A solvent trap was used to avoid sample drying out. The suspensions were prepared as 45% $TiO_2$ (wt%) in deionized water and then hand mixed for 1 minute prior loading them into the rheometer. No pre-shear was employed prior to perform any of the rheological tests, in order to keep the original agglomerate structure of the as-received powders.

The viscosity was measured as a function of shear rate, for low shear rates. The shear rate was uniformly increased and then decreased over the range of 0.3 to 10 s$^{-1}$, with a total time period of 3 minutes for both the increasing and the decreasing shear rate sweeps. Oscillatory stress sweeps were performed from 1 to 50.0 Pa in logarithmic increments, at an oscillatory frequency of 1 Hz. G' and G'' were monitored for each applied oscillatory stress pulse and recorded to further evaluate the yield stress. To investigate agglomerate breakdown, a modified freezing technique was applied, as follows: samples were taken from the suspension before and after the yield point, immersed in liquid nitrogen and then submitted to FESEM analysis. Subsequently, in order to investigate agglomerate breakdown and post-shear recovery of the aggregates, the suspensions were subjected to time sweep tests for 10 minute duration with oscillatory pulses of a fixed stress value applied at an oscillatory frequency of 1 Hz with a 3 second delay time. The applied stress values have been chosen from low values, below the yield point of each suspension, up to higher values, after the yield stress value determined for each powder. The G' buildup values were recorded as a function of time. After each time sweep test was finished, a subsequent oscillatory stress sweep was performed, as well, to determine and compare the corresponding each yield stress value with that for the suspension of the pristine powder.

The electrokinetic properties of the powders, i.e. the isoelectric point, have been measured with a Brookhaven ZetaPals instrument using the Huckel model, using diluted suspensions of TiO$_2$ (0.1g/L) in 10$^{-3}$ mol/L KCl. The pH was adjusted using solutions of either 10$^{-3}$ mol/L HCl or 10$^{-3}$ mol/L KOH, the pH range being 3-10.

Particle size distributions of diluted dispersed samples were measured with Zetasizer instrument (Malvern Instruments Ltd. UK).

FE-SEM secondary electron images of the frozen suspensions have been recorded on a LEO (Zeiss) 982 instrument at 5kV, different magnifications being used, only x10000 being shown further in the paper. Samples were prepared for FE-SEM analysis on aluminum stubs coated with carbon and consisted of frozen drops of the suspensions employed for rheological studies, taken before and after performing the stress sweep and time sweep tests, respectively, deposited on the stubs and vacuum dried. This way, changes in the aggregate structure of the suspension after applying of a certain stress could be observed and compared to the pristine powders aggregate structure.


RESULTS AND DISCUSSION

Viscosity plots as a function of shear rate for both titania samples are displayed in Figure 1. The suspensions exhibit pseudoplastic or shear thinning behavior, revealing that the particle aggregates in the suspensions were broken down into smaller flow units by increased applied shear, therefore the resistance to flow was reduced, leading to a lower viscosity as shear rate increased. Since the mechanism of shear thinning of a dispersion is a result of a reduction in the effective volume of the disperse phase caused by deflocculation of the particles upon shearing[18], the viscosity data revealed that both suspensions were in fact flocculated. In addition, it should be noted that hysteresis areas have been identified for both suspensions (Figure 1), which is a clear indication that both suspensions exhibit some thixotropy.

The rheological properties of dispersions are governed by the microstructure of the suspensions. In these systems, the solid particles are relatively small and the interparticle forces are significantly pronounced so as to influence the microstructure, the state of aggregation of the dispersion and the rheology of the system[19].

Also, recent studies on metal oxide suspensions have shown that colloidal and surface chemistry is an important factor in determining the rheological properties of concentrated colloidal suspensions[20]. The two main forces are the van der Waals attractive forces which originate from fluctuating dipoles due to the presence of outer electrons on the interacting particles, and the electrostatic repulsion forces due to the presence of charges which are, in the present case, electrostatic charges resulting from sulfate ion adsorption on the particle. For a given system, the van der Waals forces are constant. Conversely, the electrostatic forces, in the present case, will vary with the surface charge density of the suspended particles. Figure 2 shows the zeta potential variation of the two powders over a pH range, from 3 to 10. The residual sulfate present on the titania surface influenced the electrokinetic properties of the powders' surface. It is observed that higher sulfate content shifts the isoelectric point (point of zero net surface charge) towards lower pH values. In addition, a more negatively charged surface was identified for the high sulfate titania sample (Figure 2).

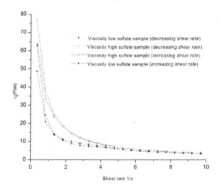

Figure 1. Viscosity variation with shear rate. for low and high sulfate titanias.

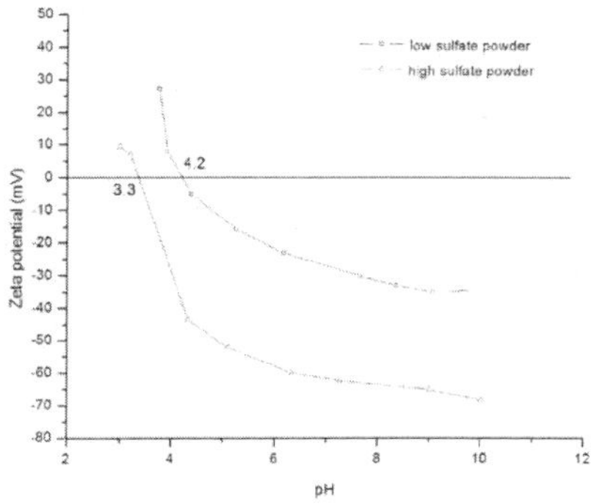

Figure 2. Zeta potential variation as. a function of pH.

Studies conducted on dispersions of titanium dioxide by previous researchers [8], found that for titanium dioxide dispersions, a change in particle size leads to totally different rheological behavior, due to a transition from an electrostatically stabilized to a sterically stabilized dispersion were observed. Yang et al.[20] found that viscosity of titanium oxide with narrow particle size distribution is higher than those of particles of broad size distribution at the same solids loading. Figure 3 presents the dynamic light scattering particle size distribution of the two powders investigated. From the number distribution shown in Figure 3, it appears that titania with a lower quantity of residual sulfate has a bi-modal particle size distribution, most of the aggregates being 170 nm, but larger ones (700 nm) are also found, while the high residual sulfate powder displays only aggregates of 700 nm. According to the zeta potential data (Figure 2), the high sulfate titania has a higher surface charge compared to the low sulfate titania, Therefore their particles should be well-dispersed in water because of the electrostatic repulsion between particles and the suspension should exhibit a low viscosity and also, smaller size of the aggregates should be expected. However, at high ionic concentration, many counterions are absorbed in the Stern layer on the particle surface. The electrostatic potential at the Helmholtz plane lowers, resulting in thin diffusion layers. Therefore, particles flocculate each other and the suspension exhibits a high shear viscosity. Conversely, a low ionic strength favors the dispersed state. The theory is in good agreement with the particle size and viscosity results, high sulfate titania exhibiting a higher viscosity at lower shear rates. The viscosity plots (Figure 1), show clearly that going up in shear rate, the initial viscosity value for the high sulfate titania is slightly higher than for the low sulfate titania. The viscosity curve recorded by going down in shear rate for the high sulfate titania crosses the going up curve recorded for the low titania at the starting viscosity value for that sample. Consequently, it appears that the flocculates that are present in the high sulfate powder have been broken down into smaller units which should be very close in size to those present in low sulfate titania.

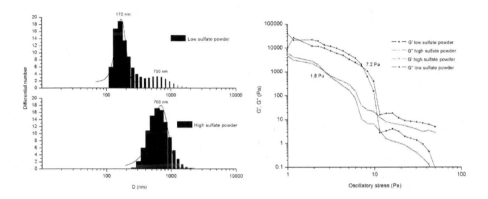

Figure 3. DLS particle size distribution.          Figure 4. Stress sweep curves of pristine powders.

Stress rheometry of the individual powders is presented in Figure 4. The suspension yield stress was determined as the crossover point between G' and G'' recorded by performing a stress sweep test. The term referred to here as 'yield stress' determined from the stress sweep is more commonly referred to as the limit of linearity. The yield stress is usually defined explicitly as the stress value corresponding to a transition from elastic solid-like behavior to viscous fluid behavior. Since the measure of these states is believed to be reflected by measurement of G' and G'', the crossover point of these two variables is argued to signify this transition. A dependence of sulfate on elasticity and yield stresses has been observed, according to Figure 4. For the titania powders investigated varying levels of interaction between the aggregates have been observed.

The lower sulfate titania seems to exhibit a greater degree of interconnectivity and elasticity, the determined yield stress being higher than for high sulfate sample. The extremely low yield point observed for the high sulfate powder suspension suggests more random interparticle interactions. Also, as mentioned before, the sulfate appears to alter the size of the effective particle. Furthermore, it can be mentioned that the high sulfate titania exhibits a weak overall elasticity which can be due to a high degree of flocculation present in the corresponding suspension. The corresponding G' values in the linear elastic regimes show that a lower stress for fracture the aggregates into free-flowing constituents. Hence, those flocculates appear to be easily broken into smaller units.

The reduction of these systems to free-flowing units from highly aggregated systems is mostly deduced and rarely visually observed. In an attempt to visually distinguish these states, samples of the suspension both in the linear elastic regime (prior to yield) and the flow regime (immediately upon yield) were taken and immediately immersed in liquid nitrogen.

The micrographs in Figures 5 and 6 appear to show a densely aggregated structure prior to yield. For the high sulfate powder, the aggregation appears to be higher than for the low sulfate titania, as already assessed by laser particle size distribution. This is seen by a large amount of what appears to be particle-particle contacts clustered into a dense assemblage.

Figure 5. FE-SEM of the pristine low sulfate Titania.     Figure 6. FE-SEM of the pristine high sulfate Titania.

By contrast, Figures 7 and 8 taken on the suspensions after yield point appear to show a greater amount of individual free-flowing units up to 1 μm in diameter. There appears to be a reduced amount of particle-particle contacts and a greater presence of individual units. The suggestion based on these micrographs is that stress-controlled rheometry results in the rupture of the network structure of the particulate suspension.

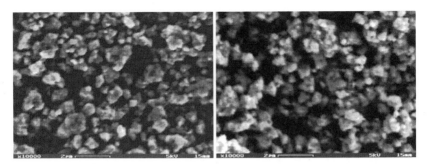

Figure 7. FESEM of the low sulfate titania after yield point.

Figure 8. FESEM of the low sulfate titania after yield point.

Time sweeps were performed on titania suspensions to determine the buildup of elasticity after a pre-shearing strain was induced. The results are presented in Figures 9 and 10. The shear stresses employed have been chosen in such a way that the suspension behavior would be still in the elastic region (values lower than the yield stress determined for each pristine powder) and also higher stresses have been employed, so the suspensions would be "flowing".

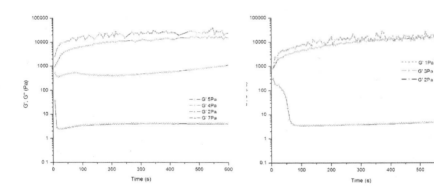

Figure 9. Time sweeps curves for low sulfate titania.

Figure 10. Time sweeps curves for high sulfate titania.

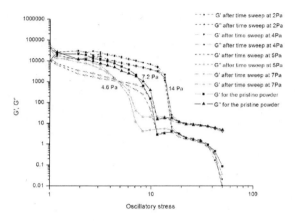

Figure 11. Stress sweep tests performed after time sweep tests for low sufate titania.

According to Figure 9, to preserve the elasticity regime, time sweeps for the low sulfate titania have been performed at 2 and 4 Pa, respectively. The higher values 5 and 7 Pa have been employed to assess aggregation after applying stresses higher than yield stress. For the high sulfate titania, pre-shearing was applied at 1, 2 Pa and 3Pa respectively. After each time sweep test was performed, a sample of suspension was frozen immediately in liquid nitrogen. Also, a **stress sweep test** was performed on the suspensions after **each time sweep test**.

Figures 9 and 10 show that elastic modulus (G') reaches a steady-state value after few tens of seconds after being pre-sheared at stresses below the yield point, which means that both systems regained their elasticity. Applying stresses higher than the yield point, (5 and 7Pa for the low sulfate powder and 3Pa for the high sulfate powder, respectively) the elasticity was not regained for any of the suspensions investigated. Figures 11 and 12 show the stress sweep curves obtained using the suspensions after each time sweep test, along with the stress sweep curves for suspensions of the pristine powders, already presented comparatively in Figure 4.

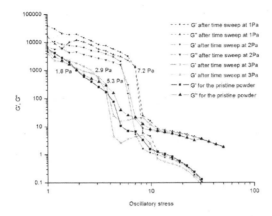

Figure 12. Stress sweep tests performed after time sweep tests for high sufate titania.

For the suspension prepared with low sulfate titania, the same value of yield stress was identified after time sweeps performed at stresses below the yield stress point of the pristine suspension. Therefore, regardless of the stress applied to the suspension within the elastic region, it appears that the aggregate structure after recovery is the same for both values employed. However, the yield stress value is higher when compared to the value of the pristine suspension. Figure 13 and 14 show the images of the low sulfate titania suspensions taken after the time sweeps performed at 2 and 4 Pa, respectively. They show smaller aggregates as compared to the image of the pristine low sulfate titania powder which explains the higher yield stress values obtained on the suspension post-time sweep testing. Usually, smaller aggregates entrain water in the system and thus increase the apparent volume fraction of the suspension, leading to an overall higher yield stress.

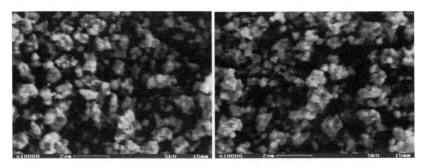

Figure 13. FE-SEM of low sulfate titania after     Figure 14. FE-SEM of low sulfate titania after
time sweep at 2Pa.                  time sweep at 4Pa.

Figure 15. FE-SEM of low sulfate titania after    Figure 16. FE-SEM of low sulfate titania after
time sweep at 5Pa.                         time sweep at 7Pa.

Therefore, it would appear that the suspension regains elasticity and the aggregate network recovery is quick and it results in smaller assemblages compared to initial powder. By applying higher stresses (5 and 7 Pa) the suspension does not regain the elasticity. The suspensions' yield stresses recorded after the time sweeps carried out at those latter pre-shear stresses are identical, but lower than for the suspension made of pristine titania. Figure 15 and 16 show the images of the low sulfate titania suspensions taken after performing the time sweeps at 5 and 7 Pa, respectively. They show a higher aggregated structure, but the loose in elasticity might be due to the irreversible rearrangement of the flowing constituents of the aggregates. A low yield stress value suggests a weak link between those aggregates after performing time sweep tests at stress values within the "flow regime".

For the suspension prepared with high sulfate titania, performing stress sweep tests on suspensions after carrying out time sweeps at different pre-shear values, led to different yield stress values. Therefore it can be assessed that the aggregate structure obtained after the G' buildup should be different. The lower the applied pre-shear, the higher yield stress value was obtained. This fact can be related to a different arrangement of the aggregates post-time sweep test, upon each pre-shear value. A higher ionic content of this high sulfate titania leads to different aggregate bond strength. Moreover, the higher the pre-shear value employed for the time sweep test, the more aggregated structure was identified. FE-SEM images were taken on frozen suspensions after each time sweep test and are presented in Figures 17-19.

Figure 17. FE-SEM of high sulfate titania after time sweep at 1Pa

Figure 18. FE-SEM of high sulfate titania after time sweep at 2Pa.

Figure 19. FE-SEM of high sulfate titania after time sweep at 3Pa.

Smaller aggregates are formed after the time sweep performed at 1 Pa, a little more aggregated structure after the test at 2Pa, which is close to that obtained at 3Pa, when the system did not regain elasticity. This pre-shear value was situated within the "flow region" of the pristine suspension.

CONCLUSIONS

Rheology of suspensions made of Titanium dioxide with two different levels of residual sulfate has found to be strongly dependent on the electrokinetic properties of the powder. The flow curves of the suspensions exhibited non-Newtonian shear thinning characteristics. A dependence of sulfate on elasticity and yield stresses has been observed. The lower sulfate titania exhibited a greater degree of interconnectivity and elasticity, the determined yield stress being higher than for high sulfate titania. For both powders, yield stresses were found to be affected by pre-shear. For low sulfate titania, yield stress after pre-shear at 2 and 4Pa > yield stress after pre-shear at 5 and 7Pa. For high sulfate titania, the yield stress values were inversely proportional to the applied pre-shear values, as follows: yield stress after pre-shear at 1Pa > yield stress after pre-shear at 2Pa > yield stress after pre-shear at 3Pa. FESEM imaging of freezed suspensions taken at different moments of rheological investigation showed varying levels of interaction between aggregates in the titania powders investigated.

ACKNOWLEDGEMENTS

The authors would like to acknowledge the financial support provided by the Ceramic and Composite Materials Research Center at Rutgers University, an NSF/IUCRC. In additions, the authors would like to thank Malvern Instruments, Ltd. for providing the instrumentation used in the testing. Lastly, the authors would like to thank Nicholas Ku and Daniel Maiorano for their technical assistance during the course of this study.

REFERENCES

[1]N. Phonthammachai, T. Chairassameewong, E. Gulari, A.M. Jamieson, S. Wongkasemjit, Structural and rheological aspect of mesoporous nanocrystalline $TiO_2$ synthesised via sol-gel process, *Microporous Mesosporous Mater* Vol 66, 2003, p. 261-171.

[2]A. Teleki, R. Wengeler, L. Wengeler, H. Nirschl, S.E. Pratsinis, Distinguishing between aggregates and agglomerates of flame-made $TiO_2$ by high-pressure dispersion, *Powder Technology* Vol 181, 2008, 292-300.

[3]W. J. Tseng, K. Lin, Rheology and colloidal structure of aqueous $TiO_2$ nanoparticle suspensions, *Materials Sci. and Engn.* A Vol 355, 2003, p. 186-192.

[4]R. Mueller, H.K. Kammler, S.E. Pratsinis, A. Vital, G. Beaucage, P. Burtscher, Non-agglomerated dry silica nanoparticles, *Powder Technology* Vol 140 (1-2), 2004, 40-48.

[5]H.K. Kammler, L. Madler, S.E. Pratsinis, Flame synthesis of nanoparticles, *Chemical Engineering and Technology* Vol 24(6), 2001, 583-596.

[6]J. Hyeon-Lee. G. Beaucage, S.E. Pratsinis, S. Vemury, Fractal analysis of flame-synthesised nanostuctured silica and titania powders using small-angle X-Ray scattering, *Langmuir* Vol 14 (20), 1998, 5751-5756.

[7]H.K. Kammler, G. Beaucage, D.J. Kohls, N. Agashe, J. Ilavsky, Monitoring simultaneously the growth of nanoparticles and aggregates by in situ ultra-small-angle X-ray scattering, *Journal of Applied Physics* Vol 97 (5) 054309.

[8]C. Saltiel, Q. Chen, S. Manickawasagam, L.S. Schadler, R.W. Siegel, M.P. Menguc, Identification of the dispersion behavior of surface treated nanoscale powders, *Journal of Nanoparticle Research* Vol 6 (1), 2004, 35-46.

[9]J. Skvarla, Evaluation of mutual interactions in binary mineral suspensions by means of electrophoretic light scattering (ELS), *Int. J. Mineral Proc.* Vol 48, 1996, 95-109.

[10]J. Farrow, L. Warren, Measurement of the size of aggregates in suspension, in: B. Dobias (Ed), Coagulation and Flocculants: Theory and Applications, Marcel Dekker, New York, 1993. p. 391-426.

[11]W. McLaughlin, J.L. White, S.L. Hem, Effect of heterocoagulation on the rheology of suspensions containing aluminium hydroxycarbonate and magnesium hydroxide, *J. Colloid Interface Sci.* Vol 167, 1994, 74-79.

[12]W.R. Richmond, R.L. Jones, P.D. Fawell, The relationship between particle aggregation and rheology in mixed silica-titania suspensions, *Chemical Engineering Journal* Vol 71, 1998, 67-75.

[13]D.W. Maiorano, N. Venugopal, R. A. Haber, The effect of the soluble sulfate concentration on the rheolgical behavior of nanoparticulate titania suspensions, *J. Ceramic Proc. Res.* Vol 8 (4), 2007, 266-270.

[14]T.F. Tadros, Rheology of Unstable Systems, in Industrial Rheology Lecture Notes (The Center for Professional Advancement, 1994.

[15]W.B. Russell, Review of the Role of Colloidal Forces in the Rheology of Suspensions, *J. Rheol.* Vol 24, 1980, 287.

[16]R.J. Hunter, Foundations of Colloid Science I; Clarendon Press, Oxford, 1987.

[17]P. V. Liddell, D.V. Boger, Influence of Processing on the Rheology of Titanium Dioxide Pigment Suspensions, *Ind. Eng. Chem. Res.* Vol 33, 1994, p. 2437-2442.

[18]J. Mewis, A.J.B. Spaull, Rheology of concentrated dispersions, *Adv. Colloid Interface Sci.* Vol 6, 1976, 173-200.

[19]P. Mikulasek, R.J. Wakeman, J.Q. Marchant, The influence of pH and temperature on the rheology and stability of aqueous titanium dioxide dispersions, *Chemical Engineering Journal* Vol 67, 1997, 97-102.

[20]H-G. Yang, C-Z. Li, H-C. Gu, T-N. Fang, Rheological Behavior of Titanium Dioxide Suspensions, *J. Colloid Interface Sci.* Vol 236, 2001, 96-103.

# HIERARCHICAL ASSEMBLY OF HYBRID NANOAPATITES: IMPLICATIONS FOR ORAL DRUG DELIVERY AND IMPLANT-BIOLOGICAL INTERFACES

Rajendra Kasinath and Allen Braizer
BioEngineering Initiative
Montana Tech of the University of Montana
Butte, Montana, USA

Kithva Hariram Prakash,
University of Queensland
Brisbane, Australia

Laurie Gower
University of Florida
Gainesville, Florida, USA

ABSTRACT

This work describes the modulation of crystallization and hierarchical structuring of hydroxyapatite-chitosan hybrid nano-particulates mediated by chitosan. Trace amounts of chitosan introduced at the onset of crystallization aided in the formation of assembled chains of equiaxed nano-HA with a narrow size distribution. Equiaxed nano-HA particles are not indicative of equilibrium morphology, as these crystals take on distinct acicular growth morphology along their c-axis typically exhibiting high aspect-ratio crystals. The highlight of this study, however, is that upon solvent extraction these hybrid suspensions form graded iridescent films which exhibited clear indications of assembly at multiple length scales: (1) below 50 nm, the ordering of nHA particles on self-assembled fibrils of chitosan was observed and (2) between 500-2000 nm, a higher order structure resembling a nano-porous sponge consisting of an interpenetrating composite of HA and chitosan was observed. The extent of assembly was dependent on the concentration of chitosan added during synthesis. The cross-section of the films revealed both structural and compositional grading with mainly chitosan on one surface and HA at the other. The implication of such hybrid nanostructures with multiple levels of structure, especially for drug delivery devices and implant-biological interfaces, is discussed further.

## INTRODUCTION

Recent exploration in the area of bio-mineralization has led to a greater understanding of how several organisms achieve non-equilibrium crystal morphogenesis.[1-3] Such organisms range from the $SiO_2$ molding diatoms[4-6] to sea-urchins[7, 8] and nacreous shells[9-11] which shape calcium carbonate minerals into a highly complex architecture often consisting of large non-equilibrium single-crystals. In this respect, the complex hierarchical ordering of apatite mineral in bone cannot be overstated.[12-15]

There is general agreement now that the mechanisms of bio-mineralization are enabled by two classes of proteins, existing at the site of mineralization. These are known as (1) traces of soluble acidic proteins and (2) a more abundant insoluble fraction of matrix proteins. Although the role of the insoluble matrix proteins can be directly linked to bio-mineral structures evident in nature, as serving a compartmental or shape directing function, the presence of soluble ones are still not fully understood.[16-18] However, several articles have proposed their function to be associated with molecular control over precipitation in the local thermodynamic and kinetic locale of mineralization.[13, 19, 20] Recently others have also shown that while acidic proteins can be effective intrinsic modulators of mineralization, they can also, perhaps extend this feature extrinsically by inducing phase separation and creating a fluid-like mineral precursor which can effectively infiltrate the predetermined matrices laid down by cells.[21-26]

In line with this paradigm we have investigated and illustrate the dramatic effects of the bio-polymer chitosan, when present in trace quantities, on the precipitation and subsequent assembly of nano-hydroxyapatite (HA) particles. The potential of HA in the field of biomedicine, especially hard tissue repair, has been well described.[27-29] Chitosan is the deactylated from of chitin, which is an abundantly present macromolecule in the make up of crustacean exoskeleton (namely shrimp and crab). Both chitin and chitosan are composed of the same monomers, N-acetyl-2-amino-2-deoxy-D-glucopyranose and 2-amino-2-deoxy-D-glucopyranose units, linked through $\beta$-D-(1-4)-glycosidic bonds. Several others[30-32] have harnessed the toughness of chitin and chitosan while incorporating HA as a bioactive component, while engineering scaffolds for hard-tissue repair. In these reports, the inclusion of chitosan is typically over 50 wt. % in the bio-composites. Concomitantly, the properties of these composites are shown to be impressive as compared with just HA. However, the present work implores to show that the presence of trace quantities of chitosan (~1-2 wt. %) can dramatically change the surface properties of the nano-apatite (nHA) and lead to self-assembly in relatively large arrays. Furthermore, increasing the amount of chitosan beyond 2 wt/ % leads to the formation of hybrid and graded organic-inorganic films. Such films could be useful in bio-interface-engineering to serve as bridges between a material construct and a biological environment (cells and tissue) through the combination of both organic and inorganic materials. The interface due to its unique structure is foreseen to provide a gradual change from one surface property for physical attachment to another for biological function.

## METHODS AND MATERIALS

Nano-Hydroxyapatite (nHA) was synthesized at 96°C by reacting calcium hydroxide-$Ca(OH)_2$ and ortho-phosphoric acid-$H_3PO_4$ (both chemicals from Sigma, Aldrich). Granulated chitosan (~80 % deactylated, from Sigma Aldrich) was dissolved in 1% acetic acid and added during HA synthesis in a controlled manner at the onset of pH change for the former reaction. Changes to the particulate system were then studied as a function of chitosan concentration (0.8-15 wt. %). nHA-chitosan hybrid films were formed by drying the respective suspensions in glass Petri dishes at 40°C, in a controlled atmosphere oven. The morphology of the HA and composite particles was observed by TEM (JEOL 2010) and structure confirmed by XRD (Shimadzu Lab 6000X). Preliminary quantification of the structure was also carried out by Rietveld full profile fitting using the Rietquan 2.3 software, the details of which are reported elsewhere.[33, 34] The morphology of the composite films was observed by SEM and FESEM (JEOL 5410 and JSM6340 respectively). The bottom surface of the films was observed by sticking the carbon tape on the film and then lifting off the tape.

The hybrid films were tested for their propensity as oral drug carrying vehicles. Insulin (Humilin® , Eli Lilly, USA) was loaded in the hybrid film by soaking 25 mg of mildly crushed hybrid films in 1 ml of insulin (100 IU/ml) at 4°C. Soaking experiments were carried out in a refrigerator over-night to maximize the degree of absorption and condensation of the drug. After soaking, the un-absorbed insulin (S1) was extracted by centrifugation of the soaked carriers. The insulin-loaded films were then dried at 4°C overnight. To the dried particles 0.5 ml of sodium alginate (2 %) followed by 0.5 ml of calcium chloride (1.5 %) was added and lightly stirred for 30 minutes, to ensure enough time for encapsulation. The suspension was then centrifuged and the supernatant (S2) was removed. The carrier was then re-suspended in distilled water and gently agitated to remove unencapsulated insulin and re-centrifuged to remove and quantify the insulin concentration (S3) in supernatant. The particles were then dried overnight at 4°C. The insulin content in the 3 supernatants (S1, S2 and S3) was measured at 271 nm using UV-Vis spectrophotometry employing suitable standards. The quantification of the insulin concentration of the supernatants (S1 + S2 + S3) yielded a relatively large loading in the carriers of about 700 IU of insulin per gram. It has to be noted that while method is a good estimate of the drug loading actual values could be slightly lower due to handling error. The encapsulated carriers (~25 mg) were each immersed in respective intestinal and gastric simulated fluids (SIF and SGF). SIF and SGF were prepared according to protocol USP-24 for SIF and SGF. A magnetic stirrer was set to 40 rpm and the dissolution media was maintained at 37°C. At 10 minute intervals, 1ml of the respective fluid was withdrawn and 1 ml of the respective media was replaced. Insulin concentration was determined by measuring absorbance and comparing to insulin standards in SIF and SGF. Release profiles were then calculated by comparing the measured release as a percentage of the loaded.

RESULTS AND DISCUSSION

The presence of chitosan in the suspension during synthesis aided in the formation of smaller nHA particles compared to its absence. TEM studies of the suspensions revealed that the particles were also better dispersed as compared to HA suspensions without chitosan (Fig. 1). The equiaxed crystalline particles had an average size of about 40±10 nm depending on the amount of chitosan added during processing.

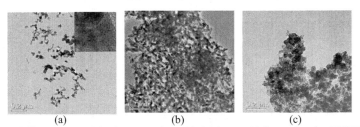

| (a) | (b) | (c) |

Figure 1 HA nano-particles in suspension with (a) 2.5 wt. % (inset showing HRTEM image), (b) 15 wt. % chitosan and (c) without chitosan addition during synthesis.

The nHA-chitosan (nHAC) suspensions appeared well dispersed immediately after synthesis. The settling time was considerably longer for these binary suspensions in contrast to pure HA suspensions. The nHA suspension settled within half an hour while the nHAC suspensions were more stable settling after longer times. In the range of 3-5 wt. % chitosan the settling time increases to a week. At higher concentration of chitosan (15 wt. %) the settling time increased to a month. Upon drying the suspensions translucent bluish films were observed, which were then peeled off by the method stated earlier. SEM revealed a self-assembled layer of nHA particles, the extent of which was dependent on the concentration of chitosan added during synthesis (Fig. 2). On the contrary, films formed from just nHA did not show arrangement and appeared aggregated (Fig. 2a). The degree of assembly was evident for chitosan content as low as 0.8 and improved up to 2.5 wt. %. At 5 wt. % the assembly started to decrease as seen by a lower degree of order and arrangement. At about 15 wt. % no specific ordering could be seen and the surface was made up of chitosan embedded with nHA (confirmed by EDX). These experiments, done in triplicate, confirmed that self-assembly was indeed dependent on chitosan concentration and that it was optimal on a glass substrate at about 2 wt. %.

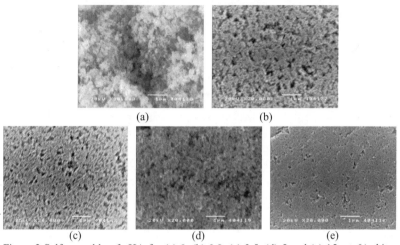

(a)          (b)

(c)        (d)        (e)

Figure 2 Self-assembly of nHA for (a) 0, (b) 0.8, (c) 2.5, (d) 5 and (e) 15 wt. % chitosan added during synthesis.

Ordering is a likely consequence of the dissolved chitosan molecules absorbed onto the nano-particles in solution. This would decrease inter-particle interaction (manifested as an increase in the settling time or defloculation). This would be a stable state determined by the balance of the attractive forces to minimize the surface energy of the system and the repulsive electrostatic forces between particle surfaces. As the chitosan concentration is increased further (>5 wt. %) there would also be an increase in viscosity of the solution and a higher interaction potential between chitosan molecules. Upon drying this stable configuration persisted as particles assumed positions relative to each other due to the steric effects imposed by the chitosan molecules such that the overall energy of the system is minimized. This is initially apparent as the amount of chitosan increased and the degree of ordering increases as well. However, beyond a certain chitosan concentration (~ 3 wt. %), the probability of chitosan chain interactions increases and leads to the overall decrease in the preferential assembly of nHA particles. XRD of intact and crushed chitosan composite films is shown in Fig. 3. The films showed higher (002) orientation as compared to the crushed powders. This orientation is the same for HA needles precipitated from neat wet chemistry. However, it is interesting to note that the arrangement of these particles on a surface also took place in the same direction. This implied that the chitosan molecules adhered or interacted predominantly with the (002) planes of nHA. This orientation was confirmed by TEM selected area diffraction (Fig. 4b), which revealed ring patterns that correspond closely to the overlapping d-spacing of (211), (112) and (300) planes. There were however only spots for (002) and (004) planes and not rings confirming that the nHA preferentially assembled on their (002) face (Fig. 4b), but were randomly rotated with respect to this plane. Phase identification of the films was carried out by XRD and subsequently the digitized XRD scans were refined by the aid of Rietveld software to determine structural details. The overall structure of HA was preserved after synthesis but the inclusion of chitosan during synthesis induced lattice parameter changes in HA. The trend (not shown here) revealed that the increase in chitosan concentration during synthesis caused a slight increase in both a and c-axis of the HA unit cell. At present we do not fully understand why this happens and so are looking at this phenomenon in greater detail. The phase quantification also revealed that there was some amount of brushite present in the films. When the XRD was carried out on only the nano-HA particle after centrifuging and washing there was no brushite detected. The fact that brushite was only observed in the films could imply that it formed during the drying stage. This finding suggested that excess chitosan molecules could be initially chelating calcium and phosphate ions in solution, and releasing them during drying.

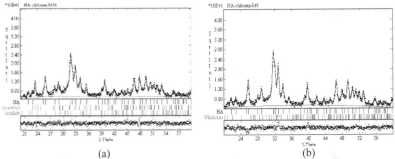

(a)           (b)

Figure 3 XRD of (a) crushed films and (b) filtered HA-chitosan particles.

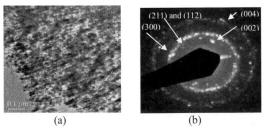

(a)           (b)

Figure 4 (a) TEM of particles assembled onto Cu-grid and (b) selected area diffraction (SAED) of the assembled layer.

Cross-sectional FESEM micrographs (Fig. 5) revealed that the films were gradually graded with different amount of chitosan/HA from the surface (mainly chitosan) to the substrate interface (mainly HA). Such a composite film implied that a film that has both organic and inorganic components can be intimately incorporated into an interpenetrant composite through a self-assembly process. Potential applications for such films include multifunctional and impervious coatings for dental implants and as cell seeding templates on in-organic surfaces. Our preliminary cell-interaction studies with osteoblasts (not shown here) indicated that this composite film is stable and supports cell growth and proliferation. Even more interestingly the study revealed that the films could specifically direct the growth and locomotion of these cells.

Figure 5 Organic-inorganic graded structure of dried HA-chitosan hybrid films

It is foreseeable that the nature of this hybrid film due to its graded structure could lead to novel drug carrying constructs. We have envisioned one such carrier system and have started carrying out preliminary studies on the propensity of these hybrid films to serve as oral drug-delivery carriers. Fig. 6 shows the schematic of a conceptualized oral drug-carrier system with several modifications to chitosan. We believe that these films, composed of several levels of structure, could be made to serve as an efficient carrier for an oral dose of peptides. The main reasons for this include: (1) the use of nHA particles to mediate and enhance cellular drug uptake and transcytosis through nonspecific means and (2) the storage of a large payload of the drug in the nanoporous architecture of the films. As illustrated in the schematic, the chitosan component in the films also enables modifications to include carbamated and thiolated residues to enhance and tailor dissolution and mucoadhesive properties of these carriers.

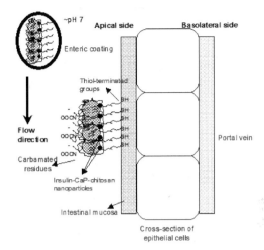

Figure 6 Conceptualized schematic of the multi-strategic nHA-chitosan carrier system for oral peptide delivery.

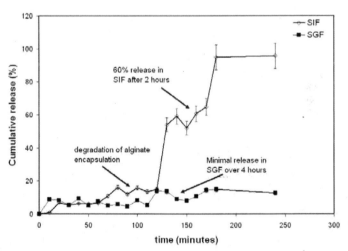

Figure 7 Release profile from the insulin loaded, encapsulated carriers in SIF (open diamonds) and SGF (closed squares)

Fig. 7 shows the release profile of insulin in SIF and SGF up to 4 hours in our proof-of-concept study, documenting the release capability of this hierarchically nanostructured carrier. The results indicated that in SGF the release maintained a level of about 10-15 % even up to 4 hours. While the release would have been diminished in SGF due to the calcium aliginate encapsulation, in SIF the higher pH causes hydrogel swelling and release was noticeable after about 1 hour. This continued to rise to about 60 % of the calculated insulin loading after 2-3 hours and reached close to 100 % release after 4 hours. The results were promising considering they were preliminary trials, and not extensively designed or fully optimized. Even so the trends did not follow simple release kinetics and seemed to be a function of several characteristics of the hybrid sponge. At first glance there seems to be release taking place due to at least two contributing kinetic factors. For now we think this release profile can be rationalized by considering the release of insulin bound with (1) nano-particles and (2) the hybrid nano-sponge architecture. At present, we are also studying the changes to the release profile as a function of chitosan molecular weight, degree of deactylation and inclusion of thiolated (for potential mucoadhesion on the intestinal epithelium) and carbamide ammonium salts (for permeation enhancement through epithelial tight junctions) into the chitosan backbone. Our ultimate aim is to study the delivery performance of this nanostructured sponge *in vivo* along with the immobilization of protease inhibitors as part of our combined delivery strategy.

CONCLUSIONS

Equiaxed, 30-40 nm HA have been synthesized out in the presence of trace quantities of chitosan. Characterization of the hybrid suspensions revealed that chitosan effectively modulated the size and assembly of crystalline nHA onto itself. The drying of the hybrid suspension formed hybrid and graded nanostructured films incorporated with self-assembled nHA particles. The proof-of-concept of using an organic-inorganic nanostructured film consisting of non-toxic and degradable materials to deliver a model peptide drug (insulin) was presented. Our preliminary results seem promising showing a dual release profile in non-cellular *in vitro* experiments using SBF and SIF to test our hypothesis. However, it is envisioned that with the versatility of chitosan to include thiolated and carbamated groups to enhance muco-adhesion and epithelial permeation, and grafting protease inhibitors to increase drug bioavailablity, the performance of this novel carrier can be substantially improved and optimized *in vivo*.

REFERENCES

[1]S. Mann, Molecular tectonics in biomineralization and biomimetic materials chemistry, *Nature*, **365**, 499-505 (1993).
[2]H. A. Lowenstam and S. Weiner, On biomineralization. 1989, New York: Oxford University Press. ix, 324.
[3]A. P. Jackson, J. F. V. Vincent, and R. M. Turner, The Mechanical Design of Nacre, *Proceedings of the Royal Society of London Series B-Biological Sciences*, **234**, 415-& (1988).
[4]N. Kroger, et al., Species-specific polyamines from diatoms control silica morphology, *Proceedings of the National Academy of Sciences of the United States of America*, **97**, 14133-14138 (2000).
[5]N. Kroger, R. Deutzmann, and M. Sumper, Polycationic peptides from diatom biosilica that direct silica nanosphere formation, *Science*, **286**, 1129-1132 (1999).
[6]S. Scala and C. Bowler, Molecular insights into the novel aspects of diatom biology, *Cellular and Molecular Life Sciences*, **58**, 1666-1673 (2001).
[7]F. H. Wilt, Matrix and mineral in the sea urchin larval skeleton, *Journal of Structural Biology*, **126**, 216-226 (1999).

[8]G. L. Decker, J. B. Morrill, and W. J. Lennarz, Characterization of Sea-Urchin Primary Mesenchyme Cells and Spicules During Biomineralization Invitro, *Development*, **101**, 297-312 (1987).

[9]J. F. Vincent, Structural Biomaterials. Revised ed. 1990, Princeton: Princeton University Press. 256 p.

[10]S. A. B. Wainwright, W.D.; Currey, J.D.; Gosline, J.M., Mechanical Designs in Organisms. 1976, Princeton, NJ: Princeton University Press.

[11]A. P. Jackson, J. F. V. Vincent, and R. M. Turner, A Physical Model of Nacre, *Composites Science and Technology*, **36**, 255-266 (1989).

[12]M. J. Glimcher, Molecular Biology of Mineralized Tissues with Particular Reference to Bone, *Reviews of Modern Physics*, **31**, 359-393 (1959).

[13]M. J. Glimcher, Recent studies of the mineral phase in bone and its possible linkage to the organic matrix by protein-ligand phosphate bonds, *Philos. Trans. R. Soc. London Ser. B*, **304**, 479-508 (1984).

[14]S. Weiner and W. Traub, Organization of Hydroxyapatite Crystals within Collagen Fibrils, *Febs Letters*, **206**, 262-266 (1986).

[15]S. Weiner and H. D. Wagner, The material bone: Structure mechanical function relations, *Annual Review of Materials Science*, **28**, 271-298 (1998).

[16]M. J. Glimcher, Mechanism of Calcification - Role of Collagen Fibrils and Collagen Phosphoprotein Complexes Invitro and Invivo, *Anatomical Record*, **224**, 139-153 (1989).

[17]J. X. Zhu, et al., Temporal and spatial gene expression of major bone extracellular matrix molecules during embryonic mandibular osteogenesis in rats, *Histochemical Journal*, **33**, 25-35 (2001).

[18]M. T. Dimuzio and A. Veis, Phosphophoryns - Major Non-Collagenous Proteins of Rat Incisor Dentin, *Calcified Tissue Research*, **25**, 169-178 (1978).

[19]L. Addadi and S. Weiner, *Stereochemical and structural relations between macromolecules and crystals in biomineralization*, in *Biomineralization - Chemical and Biochemical Perspectives*, S. Mann, J. Webb, and R.J.P. Williams, Editors. 1989, VCH Pub.: N. Y. p. 133-156.

[20]J. D. Termine, et al., Osteonectin, a Bone-Specific Protein Linking Mineral to Collagen, *Cell*, **26**, 99-105 (1981).

[21]L. A. Gower, *The Influence of Polyaspartate Additive on the Growth and Morphology of Calcium Carbonate Crystals*, in *Polymer Science & Engineering*. 1997, University of Massachusetts at Amherst. p. 119.

[22]L. A. Gower, Morphological control in biomineralization - Is it simpler than we thought?, *Abstracts of Papers of the American Chemical Society*, **217**, 040-BTEC (1999).

[23]M. J. Olszta, E. P. Douglas, and L. B. Gower, Scanning electron microscopic analysis of the mineralization of type I collagen via a polymer-induced liquid-precursor (PILP) process, *Calcified Tissue International*, **72**, 583-591 (2003).

[24]M. J. Olszta, et al., A new paradigm for biomineral formation: Mineralization via an amorphous liquid-phase precursor, *Connective Tissue Research*, **44**, 326-334 (2003).

[25]M. J. Olszta, M. Sivakumar, and L. B. Gower, Biomimetic bone via a polymer-induced liquid-precursor (PILP) process, *Science (in preparation)*, #, (2004).

[26]M. J. Olszta, et al., Nanofibrous calcite synthesized via a solution-precursor-solid mechanism, *Chemistry of Materials*, **16**, 2355-2362 (2004).

[27]H. Aoki, Medical Applications of Hydoxyapatite. 1994, St. Louis: Ishiyaku EuroAmerica.

[28]W. Bonfield and K. E. Tanner, Biomaterials - A new generation, *Materials World*, **5**, 18-20 (1997).

[29]W. Suchanek and M. Yoshimura, Processing and properties of hydroxyapatite-based biomaterials for use as hard tissue replacement implants, *Journal of Materials Research*, **13**, 94-117 (1998).

[30]I. Yamaguchi, et al., *Preparation and mechanical properties of chitosan/hydroxyapatite nanocomposites*, in *Bioceramics*. 2000. p. 673-676.

[31]I. Yamaguchi, et al., Preparation and microstructure analysis of chitosan/hydroxyapatite nanocomposites, *Journal of Biomedical Materials Research*, **55**, 20-27 (2001).

[32]H. K. Varma, et al., Porous calcium phosphate coating over phosphorylated chitosan film by a biomimetic method, *Biomaterials*, **20**, 879-884 (1999).

[33]R. Kumar, et al., Microstructure and mechanical properties of spark plasma sintered zirconia-hydroxyapatite nano-composite powders, *Acta Materialia*, **53**, 2327-2335 (2005).

[34]R. Kumar, P. Cheang, and K. A. Khor, Phase composition and heat of crystallisation of amorphous calcium phosphate in ultra-fine radio frequency suspension plasma sprayed hydroxyapatite powders, *Acta Materialia*, **52**, 1171-1181 (2004).

# Ni-B NANOLAYER EVOLUTION ON BORON CARBIDE PARTICLE SURFACES AT HIGH TEMPERATURES

Kathy Lu,* Xiaojing Zhu
Materials Science and Engineering Department
Virginia Polytechnic Institute and State University
Blacksburg, VA 24061

## ABSTRACT

Reduction study of $Ni_2O_3$ and $B_2O_3$ contained in the Ni-B nanolayer on $B_4C$ particle surfaces was carried out. For the as-coated $B_4C$ particles, the nanolayer contains $Ni_2O_3$, $B_2O_3$, and possibly amorphous B. After 400°C thermal treatment in $H_2$-Ar atmosphere, $Ni_2O_3$ is reduced to Ni while $B_2O_3$ remains. The nanolayer morphology is maintained and the particles demonstrate magnetism. As the thermal treatment temperature is further increased to 550°C and above, $B_2O_3$ is reduced to B. Subsequently, B reacts with Ni and forms $Ni_2B$. Along with the composition change, the nanolayer disappears and evolves into nanoparticles. Thermal treatment temperature increase to 800-900°C only causes $Ni_2B$ particle size growth but does not fundamentally change the composition or phase.

## INTRODUCTION

In recent years, efforts on electroless coating have shifted to coating small size cores instead of substrates. The coating layer has also evolved from micron thick to nanometer thick. A most noticeable trend is producing particles or nanoscale shells on particle surfaces. For example, Ni particles have been coated onto SiC nanoparticles.[1] Ni-P nanoparticles have been coated onto carbon nanotubes.[2] In our prior work, Ni-B nanolayer was coated onto $B_4C$ particle surfaces.[3,4] The Ni-B nanolayer thickness, morphology, and composition can be controlled by varying the electroless coating conditions.

However, one resulting issue from the electroless coating process is the oxide formation in the metallic layer. This is especially a problem when the metal layer, such as Ni, is not very inert. For the Ni-B nanolayer coated $B_4C$ particles, X-ray photoelectron spectroscopy (XPS) study detected the presence of $Ni_2O_3$ and $B_2O_3$ in the Ni-B nanolayer.[3,4] Ni-B coating on copper substrate and Ni-Co-W-B powder synthesis by similar chemical reduction methods showed the same issue.[5,6] Reduction of the resulting oxides on the core particles is necessary before the composite particles can be processed into novel materials.

NiO reduction is being actively studied. In the temperature range of 400-600°C, NiO reduction rate increases with temperature and hydrogen pressure without observable sintering. Depending on the heating rate, NiO reduction can initiate at as low as 260°C. [7,8,9] Hot stage X-ray diffraction (XRD) study showed that reduction of NiO and appearance of Ni metal cluster occur simultaneously in the temperature range of 175-300°C.[10] However, $Ni_2O_3$ reduction has not been well studied. This is likely because $Ni_2O_3$ is the less common oxide format of Ni than NiO. $B_2O_3$ was reduced by carbothermal process and high energy ball milling.[11,12,13] However, no study has been reported about reduction of $B_2O_3$ to B under controlled atmosphere, such as $H_2$, especially at low temperatures. In addition, all the above studies were focused on reduction of NiO or $B_2O_3$ only, which are different from reduction of co-existing $Ni_2O_3$ and $B_2O_3$ in the Ni-B nanolayer on the $B_4C$ particle surfaces. For the Ni-B nanolayer coated $B_4C$ particles, it is unknown how $Ni_2O_3$ and $B_2O_3$ reduction will occur and if Ni and B will form new species or not after oxide reduction.[14,15] Under the driving force for surface energy reduction, the stability of the Ni-B nanolayer also needs to be examined.

This study is focused on the reduction of $Ni_2O_3$ and $B_2O_3$ contained in the Ni-B nanolayer on $B_4C$ particle surfaces. Scanning electron microscopy (SEM) is used to examine the Ni-B nanolayer morphology evolution. XPS is used to study the Ni-B nanolayer composition change. XRD is used to analyze the Ni-B nanolayer phase evolution at different temperatures. A Ni-B nanolayer evolution process at different thermal treatment conditions is presented.

EXPERIMENTAL PROCEDURE

A tube furnace (Rapid Temp Model 1730-20HT, CM Furnaces, Bloomfield, NJ) was used to reduce the $Ni_2O_3$ and $B_2O_3$ in the Ni-B nanolayer that exists on the $B_4C$ particle surfaces. The reducing atmosphere was a 3:5 volume ratio mixture of $H_2$:Ar. $B_4C$ particles coated with 40-60 nm thickness Ni-B nanolayer were put on a Ni foil. The Ni foil was then put in the center of the tube furnace for $Ni_2O_3$ and $B_2O_3$ reduction. The tube furnace was first purged with Ar for 10 min to remove any residual air. After that, the furnace temperature was increased. When the peak thermal treatment temperature was lower than or at 800°C, the heating rate used was 200°C/hr. When the peak temperature was greater than 800°C, the heating rate used was 100°C/hr at temperatures higher than 800°C. The samples were kept at the targeted peak temperature for 2 hours. The furnace was then cooled down to room temperature with the same rate. Ar gas was then used to purge the furnace for 10 min to remove any residual hydrogen.

A field emission SEM (LEO 1550, Carl Zeiss MicroImaging, Inc, Thornwood, NY) was used to characterize the morphology of the Ni-B nanolayer on the $B_4C$ particle surfaces. XPS (Phi Quantera SXM-03, Physical Electronics, Inc., Chanhassen, MN) was employed to characterize the composition change of the Ni-B nanolayers. XRD (X'Pert PRO diffractometer, PANalytical B.V., EA Almelo, The Netherlands) was carried out to identify the crystalline phases on the $B_4C$ particle surfaces at as-coated condition and after different thermal treatment temperatures. The XRD experiment step size was 0.05°, the dwell time per step was 100 s, and the scan rate was $0.063°·s^{-1}$ with $CuK_\alpha$ radiation ($\lambda = 1.5406$ Å) and a Ni filter.

RESULTS AND DISCUSSION

Morphology Evolution of Ni-B Nanolayer

The as-coated Ni-B nanolayer on the $B_4C$ particle surfaces has a mesh-like structure.[3,4] Ni-B nodules randomly embedded in the mesh structure. Ni:B elemental ratio is 2.51:1 for the nanolayer. In this study, the Ni-B nanolayer coated $B_4C$ particle samples remain soft and non-agglomerated after thermal treatment at 400-900°C temperatures in the $H_2$-Ar atmosphere. However, the thermally treated samples at all temperatures are shinier. Fig. 1 shows the SEM images of the Ni-B nanolayer coated $B_4C$ samples at as-coated state and after thermal treatment at different temperatures. The as-coated sample has bright Ni-B nodules embedded in the mesh structures as mentioned. The sample treated at 400°C retains the meshed nanolayer surface morphology. However, there is a drastic Ni-B nanolayer morphology transformation when the thermal treatment temperature is increased to 550°C. At 550°C thermal treatment temperature, the meshed nanolayer structure disappears and nanoparticles appear on the $B_4C$ particle surfaces. The initial nanoparticles are very small and barely visible. As the thermal treatment temperature further increases to 800°C and 900°C, the nanoparticle sizes grow. Based on the size measurement of 50 particles on the $B_4C$ particle surfaces, the particle size is approximately 10.6 nm at 550°C. At 800°C, the particle size grows substantially to 78.1 nm. At 900°C, the particle size grows further to 82.6 nm. The particle size increase (in nm) is proportional to the logarithm of absolute thermal treatment temperature with the highest particle size growth rate around 800°C.

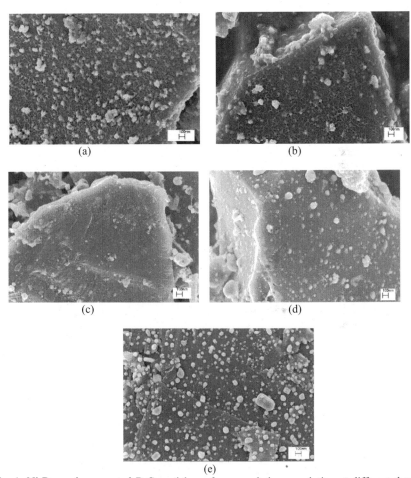

(a)

(b)

(c)

(d)

(e)

Fig. 1. Ni-B nanolayer coated $B_4C$ particle surface morphology evolution at different thermal treatment temperatures in $H_2$-Ar atmosphere: (a) as-coated, (b) 400°C, (c) 550°C, (d) 800°C, and (e) 900°C.

The morphology evolution in Fig. 1 indicates that the Ni-B nanolayer on the $B_4C$ particle surfaces is thermodynamically unstable at elevated temperatures. The surface nanolayer instability leads to break-up and reconstruction of the Ni-B coating in a manner similar to a dewetting process. Spheroidization of the Ni-B nanolayer leads to total surface energy reduction and more stable states. This Ni-B nanolayer evolution finally leads to a set of Ni-B nanoparticles on the core $B_4C$ particles. Since the original $B_4C$ particles are not ideally smooth and some smaller $B_4C$ particles are attached to the large $B_4C$ particle surfaces,[3] the Ni-B nanolayer can spheroidize on the large host $B_4C$ particle surfaces or on/around the attached, small $B_4C$ particle surfaces, both through heterogeneous nucleation process. When the Ni-B particle formation is on the large host $B_4C$ particle surfaces, nucleation occurs at randomly distributed sites, leading to the formation of the smaller Ni-B particles. When the Ni-B particle formation is on/around the attached, small $B_4C$ particle surfaces, these small $B_4C$ particles provide nucleation centers and lead to the formation of some much larger Ni-B particles even though the population of the large Ni-B particles is much smaller. The end result is a bi-modal Ni-B particle size distribution on the $B_4C$ particle surfaces with a small percent of large Ni-B particles.

Composition Evolution of Ni-B Nanolayer

In this study (Fig. 2), the as-coated $B_4C$ sample with the Ni-B nanolayer coating shows boron-related peaks at 187 eV and 191.6 eV. The Ni-B nanolayer coated $B_4C$ samples at different thermal treatment temperatures show the boron-related peaks at 187 eV and 192.5 eV. The slight left shift of the 192.5 eV peak at as-coated condition is likely due to the presence of residual $NaBH_4$ or $C_2H_2N_8$ used during the electroless coating process.[3,4] Since the binding energies of B in pure B, $B_4C$, and Ni-B alloy compound(s) overlap at 187 eV,[16, 17,18] it is impossible to determine the specific B-related species from this 187 eV XPS peak. However, it can be determined that the 187 eV energy peak is related to the B species at non-oxidized state. The shoulder-like appearance for the right side of the 187 eV peak at higher thermal treatment temperatures seems to indicate the increasing amount of B.[17] The 192.5 eV binding energy peak can be $B_2O_3$ or $Ni_xB_{100-x}O_y$ based on the literature.[18,19,20,21,] This is also the peak related to the B species at oxidized state.

From the relative height and area of the two boron related XPS peaks at 187 eV and 192.5 eV, it can be concluded that there is more B in the oxidized state than in the non-oxidized state for the as-coated sample and the 400°C thermally treated sample. As the thermal treatment temperature increases to 550°C, the XPS spectrum shows the non-oxidized B-related peak becomes more dominant than the oxidized B-related peak. This shift can be attributed to two concurrent processes. One is that the oxidized boron species start to be reduced by hydrogen at 550°C and higher temperatures to non-oxidized boron species, most likely to pure B or Ni-B alloy compound(s). The other accompanying process is the Ni-B nanolayer morphology evolution. As the oxygen in $B_2O_3$ and possibly in $Ni_xB_{100-x}O_y$ is removed, the Ni-B nanolayer becomes thinner. This thin Ni-B nanolayer is more likely to become unstable and spheroidize, as shown in Fig. 1, resulting in the $B_4C$ surface being exposed. From the comparison of the B-related peaks at different thermal treatment temperatures (Fig. 2), the samples treated at 800°C give the lowest boron related peak at ~192 eV (oxidized state) and thus are must desired for $B_2O_3$ reduction. As to the specific oxidized and non-oxidized B species, the XPS spectra for the Ni related species need to be considered. Also, XRD results will provide more information for phase identification.

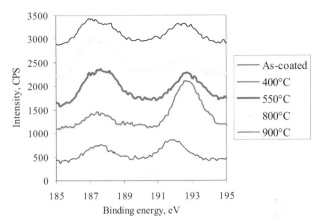

Fig. 2. XPS spectra showing boron related peaks of Ni-B nanolayer coated B₄C samples treated at different temperatures.

For the Ni related XPS spectra, the as-coated sample shows four peaks (Fig. 3). Both NiO and Ni$_2$O$_3$ have four Ni-related peaks in their respective XPS spectra. But the peak positions are different from each other.[18,22,23] For the as-coated sample the primary nickel peak is at 856 eV, which is the primary Ni 2p3/2 peak. This indicates that the Ni-related species in the Ni-B nanolayer is Ni$_2$O$_3$, not NiO.[23,24] For the Ni-B nanolayer coated B$_4$C sample thermally treated at 400°C, magnetism is detected by placing the Ni-B coated B$_4$C particles next to a permanent magnet. Also, the primary Ni 2p3/2 peak rises at 852.5 eV, which is characteristic of Ni.[23,24] The relative intensity of the four XPS peaks at 400°C is also characteristic of pure Ni. These observations show that Ni$_2$O$_3$ has been reduced to pure Ni after 400°C thermal treatment. Combining with Fig. 2, it also means that the presence of Ni$_x$B$_{100-x}$O$_y$ in the Ni-B nanolayer is negligible. Otherwise, Ni$_x$B$_{100-x}$O$_y$ should be reduced at 400°C and such reduction should lead to the decrease, not increase, of the 192.5 eV peak in the oxidized boron-related spectra. As the thermal treatment temperature is further increased to 550°C and higher, the satellite peaks for Ni become subdued. Only the peaks at 852.5 eV and 870 eV are visible. Also the peaks at these two binding energy positions become sharper and the samples under such thermal treatment conditions show no magnetism. This means Ni in the Ni-B nanolayer has substantially decreased, likely to an undetectable degree. The 852.5 eV binding energy peak is identified to be the primary Ni 2p3/2 peak from Ni$_2$B. The 870 eV binding energy peak is the Ni 2p$_{1/2}$ satellite peak for Ni$_2$B. This means as temperature increases, Ni reacts with B and forms a new species, Ni$_2$B. The B species that has participated in the Ni$_2$B formation should come from amorphous B$_2$O$_3$ or B.[16] This is likely accompanied by B$_2$O$_3$ reduction to B under H$_2$-Ar reducing atmosphere. As reported,[4] the Ni:B elemental ratio is 2.51:1 for the Ni-B nanolayer coating. With Ni$_2$B formation, most of the Ni will be consumed. This explains why the Ni peak diminishes as temperature increases. Correlating with Fig. 2, this means the 187 eV binding energy boron peak at temperatures greater than 400°C are partly contributed by the formation of Ni$_2$B.

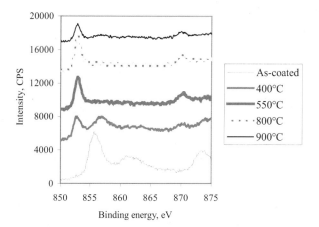

Fig. 3. XPS spectra showing nickel-related peaks for Ni-B coated $B_4C$ samples thermally treated at different temperatures.

Phase Evolution of Ni-B Nanolayer

Fig.4 shows the XRD patterns for the Ni-B nanolayer coated $B_4C$ particles with no thermal treatment, and the Ni-B nanolayer coated $B_4C$ particles at different thermal treatment conditions. The peaks from the $B_4C$ species persist at all conditions even though the relative intensity changes in comparison to the other peaks. The $Ni_2O_3$ peaks in the as-coated sample are not displayed because of the low intensity of the peaks. As the thermal treatment temperature increases to 400°C, Ni peaks appear. This is consistent with the XPS results in Fig. 3. As the temperature is further increased to 550°C and higher, the Ni peaks diminish and the $Ni_2B$ peaks appear. This means that $B_2O_3$ reduction to B occurs. With the presence of Ni, $Ni_2B$ forms by consuming Ni and B. From 550°C to 900°C thermal treatment temperatures, there are no changes in the crystalline phases for the Ni-B coated $B_4C$ particles. Based on the morphological change shown in Fig. 1, $Ni_2B$ formation is accompanied by the conversion of Ni-B nanolayer into $Ni_2B$ nanoparticles.

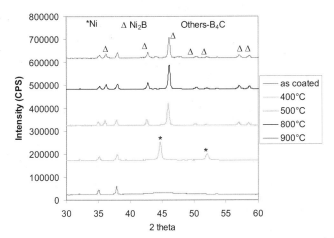

Fig. 4 XRD patterns for the as-is $B_4C$ particles, Ni-B nanolayer coated $B_4C$ particles with no thermal treatment, and Ni-B nanolayer coated $B_4C$ particles at different thermal treatment conditions.

Based on the above analyses, the Ni-B nanolayer morphology and composition evolution on the $B_4C$ particle surfaces at different thermal treatment conditions can be understood as follows. For the as-coated $B_4C$ particles, the nanolayer contains B, $Ni_2O_3$, and $B_2O_3$. At 400°C thermal treatment conditions, $Ni_2O_3$ is reduced to Ni while the $B_2O_3$ species remains. The nanolayer morphology is maintained and the particles demonstrate magnetism. As the thermal treatment temperature increases to 550°C, $B_2O_3$ is reduced to B. B reacts with Ni and forms $Ni_2B$. Simultaneously, the Ni-B nanolayer evolves into nanoparticles. Further temperature increase to 800°C-900°C only causes $Ni_2B$ particle size growth but does not fundamentally change the composition or phase.

CONCLUSIONS

Ni-B nanolayer evolution on $B_4C$ particle surfaces under $H_2$-Ar reducing atmosphere is studied. SEM is used to examine the Ni-B nanolayer morphology evolution. XPS is used to study the Ni-B nanolayer composition change. XRD is used to analyze the Ni-B nanolayer phase evolution at different temperatures. The study shows that for the as-coated $B_4C$ particles, the nanolayer contains $Ni_2O_3$, $B_2O_3$, and possibly amorphous B. At 400°C thermal treatment condition, $Ni_2O_3$ is reduced to Ni while the $B_2O_3$ species remains. The nanolayer morphology is maintained and demonstrates magnetism. As the thermal treatment temperature is increased to 550°C, $B_2O_3$ is reduced to B. B reacts with Ni and forms $Ni_2B$. Also, the Ni-B nanolayer evolves into nanoparticles. Further temperature increase to 800-900°C only causes particle size growth but does not fundamentally change the nanoparticle composition or phase.

ACKNOWLEDGMENT

The authors acknowledge the financial support from National Science Foundation under grant No. DMI-0620621.

FOOTNOTE

Member, American Ceramic Society

REFERENCES:

[1] Y. J. Chen, M. S. Cao, Q. Xu and J. Zhu, Electroless Nickel Plating on Silicon Carbide Nanoparticles, *Surf. Coat. Technol.* **172**, 90-94 (2003).

[2] F. Wang, S. Arai, K. C. Park, K. Takeuchi, Y. J. Kim and M. Endo, Synthesis of Carbon Nanotube-Supported Nickel-Phosphorus Nanoparticles by an Electroless Process, *Carbon* **44**, 1307-1310 (2006).

[3] H. Dong, X. Zhu, K. Lu, Morphology and Composition of Nickel-Boron Nanolayer Coating on $B_4C$ Particles, *J. Mater. Sci.* **43**, 4247-4256 (2008).

[4] X. Zhu, H. Dong, K. Lu, Coating Different Thickness Nickel-Boron Nanolayers onto Boron Carbide Particles, *Surf. Coat. Technol.* **202**, 2927-2934 (2007).

[5] M. V. Ivanov, E. N. Lubnin, and A. B. Drovosekov, X-Ray Photoelectron Spectroscopy of Chemically Deposited Amorphous and Crystalline Ni–B Coatings, *Prot. Met.* **39**, 133-141 (2003).

[6] W. L. Dai, M. H. Qiao, and J. F. Deng, XPS Studies on a Novel Amorphous Ni-Co-W-B Alloy Powder, *Appl. Surf. Sci.* **120**, 119-124 (1997).

[7] T. A. Utigard, M. Wu, G. Plascencia, and T. Marin, Reduction Kinetics of Goro Nickel Oxide Using Hydrogen, *Chem. Eng. Sci.* **60**, 2061-2068 (2005).

[8] M. Lorenz and M. Schulze, Reduction of Oxidized Nickel Surfaces, *Surf. Sci.* **454**, 234-239 (2000).

[9] B. Jankovic, B. Adnadevic, and S. Mentus, The Kinetic Analysis of Non-Isothermal Nickel Oxide Reduction in Hydrogen Atmosphere Using the Invariant Kinetic Parameters Method, *Thermochim Acta* **456**, 48-55 (2007).

[10] J. T. Richardson, R. Scates, and M. V. Twigg, X-ray Diffraction Study of Nickel Oxide Reduction by Hydrogen, *Appl. Catal. A* **246**, 137-150 (2003).

[11] F. Deng, H. Y. Xie, and L. Wang, Synthesis of Submicron $B_4C$ by Mechanochemical Method, *Mater. Lett.* **60**, 1771-1773 (2006).

[12] C. H. Jung, M. J. Lee and C. J. Kim, Preparation of Carbon-Free $B_4C$ Powder From $B_2O_3$ Oxide by Carbothermal Reduction Process, *Mater. Lett.* **58**, 609-614 (2004).

[13] R. Ricceri and P. Matteazzi, Mechanochemical Synthesis of Elemental Boron, *Int. J. Powder Metall.* **39**, 48-52 (2003).

[14] K. I. Portnoi and V. M. Romashov, Binary Constitution Diagrams of Systems Composed of Various Elements and Boron - A Review, *Soviet Powder Metall. Metal Ceram.* **11**, 378-384 (1972).

[15] K. E. Spear, Correlations and Predictions of Metal-Boron Phase Equilibria, Applications of Phase Diagrams in Metallurgy and Ceramics, *NBS Special Pub.* **2**, 744-762 (1978).

[16] D. N. Hendrickson, J. M. Hollander and W. L. Jolly, Core-Electron Binding Energies for Compounds of Boron, Carbon, and Chromium, *Inorg. Chem.* **9**, 612-615 (1970).

[17]L. Chen, T. Goto, T. Hirai and T. Amano, State of Boron in Chemical Vapour-Deposited SiC-B Composite Powders, *J. Mater. Sci. Lett.* **9**, 997-999 (1990).

[18]J. A. Schreifels, P. C. Maybury and W. E. Swartz, X-Ray Photoelectron Spectroscopy of Nickel Boride Catalysts: Correlation of Surface States with Reaction Products in the Hydrogenation of Acrylonitrile, *J. Catal.* **65**, 195-206 (1980).

[19]A. R. Burke, C. R. Brown, W. C. Bowling, J. E. Glaub, D. Kapsch, C. M. Love, R. B. Whitaker, W. E. Moddeman, Ignition Mechanism of the Titanium-Boron Pyrotechnic Mixture, *Surf. Interface Anal.* 11, 353 (1988).

[20]W. A. Brainard, D. R. Wheeler, An XPS Study of the Adherence of Refractory Carbide Silicide and Boride RF-Sputtered Wear-Resistant Coatings, *J. Vac. Sci. Technol.* 15, 1801 (1978).

[21]J. Tamaki, H. Takagaki, T. Imanaka, Surface Characterization of Amorphous Ni-B Films Treated with Oxygen, *J. Catal.* 108, 256 (1987).

[22]A. N. Mansour and C. A. Melendres, Characterization of $Ni_2O_3 \cdot 6H_2O$ by XPS, *Surf. Sci. Spectra* **3**, 263-270 (1996).

[23]C. D. Wagner, W. M. Riggs, L. E. Davis, J. F. Moulder and G. E. Muilenberg, Handbook of X-Ray Photoelectroon Spectroscopy-A Reference Book of Standard Data for Use in X-Ray Photoelectron Spectroscopy, Perkin-Elmer Corporation, Physical Electronics Division, Eden Prairie, Minnesota, 1979.

[24]Y.-X. Zheng, S.-B. Yao, and S.-M. Zhou, Study on Antioxidation of Nanosize Ni-Mo-B Amorphous Alloy, *Acta Phys.-Chin. Sin.* **20**, 1352-1356 (2004).

# Author Index